Lessons from Nanoelectronics

A New Perspective on Transport
— Part B: Quantum Transport

Lessons from Nanoscience: A Lecture Note Series

ISSN: 2301-3354

Series Editors: Mark Lundstrom and Supriyo Datta
(Purdue University, USA)

"Lessons from Nanoscience" aims to present new viewpoints that help understand, integrate, and apply recent developments in nanoscience while also using them to re-think old and familiar subjects. Some of these viewpoints may not yet be in final form, but we hope this series will provide a forum for them to evolve and develop into the textbooks of tomorrow that train and guide our students and young researchers as they turn nanoscience into nanotechnology. To help communicate across disciplines, the series aims to be accessible to anyone with a bachelor's degree in science or engineering.

More information on the series as well as additional resources for each volume can be found at: http://nanohub.org/topics/LessonsfromNanoscience

---

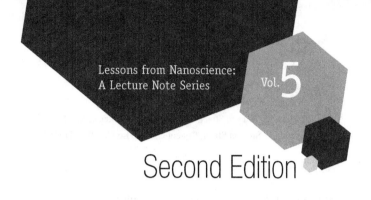

Lessons from Nanoscience:
A Lecture Note Series

Vol. 5

Second Edition

Lessons from Nanoelectronics

A New Perspective on Transport
— Part B: Quantum Transport

Supriyo Datta

Purdue University, USA

World Scientific

NEW JERSEY · LONDON · SINGAPORE · BEIJING · SHANGHAI · HONG KONG · TAIPEI · CHENNAI · TOKYO

Published by

World Scientific Publishing Co. Pte. Ltd.

5 Toh Tuck Link, Singapore 596224

USA office: 27 Warren Street, Suite 401-402, Hackensack, NJ 07601

UK office: 57 Shelton Street, Covent Garden, London WC2H 9HE

Library of Congress Cataloging-in-Publication Data
Names: Datta, Supriyo, 1954– author.
Title: Lessons from nanoelectronics : a new perspective on transport /
    Supriyo Datta, Purdue University, USA.
Other titles: Lessons from nanoscience ; v. 5.
Description: Second edition. | Singapore ; Hackensack, NJ : World Scientific, [2017]– |
    Series: Lessons from nanoscience: a lecture notes series, ISSN 2301-3354 ; vol. 5 |
    Includes bibliographical references and index.
Identifiers: LCCN 2016058007| ISBN 9789813224643 (set : alk. paper) |
    ISBN 9789813224650 (set : pbk. ; alk. paper) |
    ISBN 9789813209732 (Part A ; hardcover ; alk. paper) |
    ISBN 9813209739 (Part A ; hardcover ; alk. paper) |
    ISBN 9789813209749 (Part A; pbk. ; alk. paper) | ISBN 9813209747 (Part A ; pbk. ; alk. paper)
    ISBN 9789813224605 (Part B ; hardcover ; alk. paper) |
    ISBN 9813224606 (Part B ; hardcover ; alk. paper) |
    ISBN 9789813224612 (Part B; pbk. ; alk. paper) | ISBN 9813224614 (Part B ; pbk. ; alk. paper)
Subjects: LCSH: Nanoelectronics. | Transport theory.
Classification: LCC TK7874.84 .D37 2017 | DDC 621.381--dc23
LC record available at https://lccn.loc.gov/2016058007

Lessons from Nanoelectronics: A New Perspective on Transport (Volume 1)
ISBN: 9789814335287
ISBN: 9789814335294 (pbk)

British Library Cataloguing-in-Publication Data
A catalogue record for this book is available from the British Library.

Printed in Singapore

To my students and colleagues

and all others whose love of learning

has helped me learn

Preface

Everyone is familiar with the amazing performance of a modern smartphone, powered by a billion-plus nanotransistors, each having an active region that is barely a few hundred atoms long. I believe we also owe a major intellectual debt to the many who have made this technology possible.

This is because the same amazing technology has also led to a deeper understanding of the nature of current flow and heat dissipation on an atomic scale which I believe should be of broad relevance to the general problems of non-equilibrium statistical mechanics that pervade many different fields.

To make these lectures accessible to anyone in any branch of science or engineering, we assume very little background beyond linear algebra and differential equations. However, we will be discussing advanced concepts that should be of interest even to specialists, who are encouraged to look at my earlier books for additional technical details.

This book is based on a set of two online courses originally offered in 2012 on nanoHUB-U and more recently in 2015 on edX. In preparing the second edition we decided to split it into parts A and B entitled *Basic Concepts* and *Quantum Transport* respectively, along the lines of the two courses. Even this Second Edition represents lecture notes in unfinished form.

For ease of reference, Part B includes a list of all equations from Part A that are referred to (Appendix F). We have also included the overview chapter (Chapter 1) from Part A with essentially no modification. A list of available *video lectures* corresponding to different sections of this volume is provided upfront. I believe readers will find these useful.

Acknowledgments

The precursor to this lecture note series, namely the *Electronics from the Bottom Up* initiative on https://nanohub.org was funded by the U.S. National Science Foundation (NSF), the Intel Foundation, and Purdue University. Thanks to World Scientific Publishing Corporation and, in particular, our series editor, Zvi Ruder for joining us in this partnership.

In 2012 nanoHUB-U offered its first two online courses based on this text. We gratefully acknowledge Purdue and NSF support for this program, along with the superb team of professionals who made nanoHUB-U a reality (https://nanohub.org/u) and later helped offer these courses through edX.

A special note of thanks to Mark Lundstrom for his leadership that made it all happen and for his encouragement and advice. I am grateful to Shuvro Chowdhury for carefully going through this new edition and fixing the many errors introduced during conversion to LATEX. I also owe a lot to many students, ex-students, on-line students and colleagues for their valuable feedback and suggestions regarding these lecture notes.

Finally I would like to express my deep gratitude to all who have helped me learn, a list that includes many students, colleagues and teachers starting with the late Richard Feynman whose classic lectures on physics, I am sure, have inspired many like me and taught us the "pleasure of finding things out."

Supriyo Datta

List of Available Video Lectures Quantum Transport

This book is based on a set of two online courses originally offered in 2012 on nanoHUB-U and more recently in 2015 on edX. These courses are now available in self-paced format at nanoHUB-U (https://nanohub.org/u) along with many other unique online courses.

Additional information about this book along with questions and answers is posted at the book website.

In preparing the second edition we decided to split the book into parts A and B following the two online courses available on nanoHUB-U entitled *Fundamentals of Nanoelectronics*

Part A: Basic Concepts   Part B: Quantum Transport

Also of possible interest in this context: NEGF: A Different Perspective.

Following is a detailed list of *video lectures* available at the course website corresponding to different sections of this volume (Part B: Quantum Transport). The figures of this volume can be regenerated with these MATLAB codes which are also available in Appendix H.

In the following QTAT stands for
Quantum Transport: Atom to Transistor, Chapter 5–11, Cambridge (2005). To reproduce the figures of the book use these MATLAB codes.

Constants Used in This Book

Electronic charge	$-q$	$=$	-1.6×10^{-19} C (Coulomb)
Unit of energy	1 eV	$=$	1.6×10^{-19} J (Joule)
Boltzmann constant	k	$=$	1.38×10^{-23} J $\cdot$ K^{-1}
			~ 25 meV (at 300 K)
Planck's constant	h	$=$	6.626×10^{-34} J $\cdot$ s
Reduced Planck's constant	$\hbar = h/2\pi$	$=$	1.055×10^{-34} J $\cdot$ s
Free electron mass	m_0	$=$	9.109×10^{-31} kg

Some Symbols Used

m	Effective Mass	Kg
I	Electron Current	A (Amperes)
T	Temperature	K (Kelvin)
t	Transfer Time	s (second)
V	Electron Voltage	V (Volt)
U	Electrostatic Potential	eV
μ	Electrochemical Potential (also called Fermi level or quasi-Fermi level)	eV
μ_0	Equilibrium Electrochemical Potential	eV
R	Resistance	Ω (Ohm)
G	Conductance	S (Siemens)
$G(E)$	Conductance at 0 K with $\mu_0 = E$	S
λ	Mean Free Path for Backscattering	m
L_E	Energy Relaxation Length	m
L_{in}	Mean Path between Inelastic Scattering	m
τ_m	Momentum Relaxation time	s
$\overline{D}$	Diffusivity	$m^2 \cdot s^{-1}$
$\overline{\mu}$	Mobility	$m^2 \cdot V^{-1} \cdot s^{-1}$
ρ	Resistivity	$\Omega \cdot m$ (3D), Ω (2D), $\Omega \cdot m^{-1}$ (1D)
σ	Conductivity	$S \cdot m^{-1}$ (3D), S (2D), $S \cdot m$ (1D)
$\sigma(E)$	Conductivity at 0 K with $\mu_0 = E$	$S \cdot m^{-1}$ (3D), S (2D), $S \cdot m$ (1D)
A	Area	m^2

W	Width	m
L	Length	m
E	Energy	eV
C	Capacitance	F (Farad)
ϵ	Permittivity	$\mathrm{F \cdot m^{-1}}$
ε	Energy	eV
$f(E)$	Fermi Function	Dimensionless
$\left(-\frac{\partial f}{\partial E}\right)$	Thermal Broadening Function (TBF)	$\mathrm{eV^{-1}}$
$kT\left(-\frac{\partial f}{\partial E}\right)$	Normalized TBF	Dimensionless
$D(E)$	Density of States	$\mathrm{eV^{-1}}$
$N(E)$	Number of States with Energy $< E$ (equals number of Electrons at 0 K with $\mu_0 = E$)	Dimensionless
n	Electron Density (3D or 2D or 1D)	$\mathrm{m^{-3}}$ or $\mathrm{m^{-2}}$ or $\mathrm{m^{-1}}$
n_s	Electron Density	$\mathrm{m^{-2}}$
n_L	Electron Density	$\mathrm{m^{-1}}$
$M(E)$	Number of Channels (also called transverse modes)	Dimensionless
ν	Transfer Rate	$\mathrm{s^{-1}}$
$\gamma = \hbar\nu$	Energy Broadening	eV
$\mathbf{X}^\dagger$	Complex conjugate of transpose of matrix $\mathbf{X}$	
$\mathbf{H}$	(Matrix) Hamiltonian	eV
$\mathbf{G}^R(E)$	(Matrix) Retarded Green's function	$\mathrm{eV^{-1}}$
$\mathbf{G}^n(E)/2\pi$	(Matrix) Electron Density	$\mathrm{eV^{-1}}$, per gridpoint
$\mathbf{A}(E)/2\pi$	(Matrix) Density of States	$\mathrm{eV^{-1}}$, per gridpoint
$\mathbf{\Gamma}(E)$	(Matrix) Energy Broadening	eV
$\mathbf{G}^A(E)$ $= [\mathbf{G}^R(E)]^\dagger$	(Matrix) Advanced Green's function	$\mathrm{eV^{-1}}$
BTE	**B**oltzmann **T**ransport **E**quation	
NEGF	**N**on-**E**quilibrium **G**reen's **F**unction	
DOS	**D**ensity **O**f **S**tates	
QFL	**Q**uasi-**F**ermi **L**evel	

Contents

Chapter 1

Overview

This chapter is essentially the same as Chapter 1 from Part A. Related video lecture available at course website, Scientific Overview.

"Everyone" has a smartphone these days, and each smartphone has more than a billion transistors, making transistors more numerous than anything else we could think of. Even the proverbial ants, I am told, have been vastly outnumbered.

There are many types of transistors, but the most common one in use today is the Field Effect Transistor (FET), which is essentially a resistor consisting of a "channel" with two large contacts called the "source" and the "drain" (Fig. 1.1a).

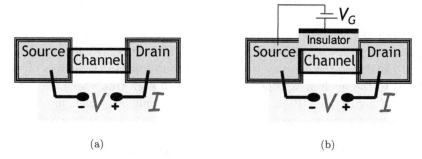

(a)   (b)

Fig. 1.1 (a) The Field Effect Transistor (FET) is essentially a resistor consisting of a channel with two large contacts called the source and the drain across which we attach the two terminals of a battery. (b) The resistance $R = V/I$ can be changed by several orders of magnitude through the gate voltage V_G.

The resistance $(R) =$ Voltage (V)/Current (I) can be switched by several orders of magnitude through the voltage V_G applied to a third terminal called the "gate" (Fig. 1.1b) typically from an "OFF" state of ~ 100 MΩ to an "ON" state of ~ 10 kΩ. Actually, the microelectronics industry uses a complementary pair of transistors such that when one changes from 100 MΩ to 10 kΩ, the other changes from 10 kΩ to 100 MΩ. Together they form an inverter whose output is the "inverse" of the input: a low input voltage creates a high output voltage while a high input voltage creates a low output voltage as shown in Fig. 1.2.

A billion such switches switching at GHz speeds (that is, once every nanosecond) enable a computer to perform all the amazing feats that we have come to take for granted. Twenty years ago computers were far less powerful, because there were "only" a million of them, switching at a slower rate as well.

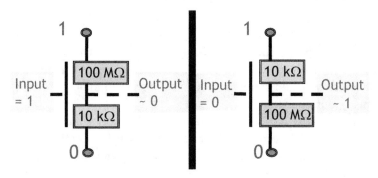

Fig. 1.2   A complementary pair of FET's form an inverter switch.

Both the increasing number and the speed of transistors are consequences of their ever-shrinking size and it is this continuing miniaturization that has driven the industry from the first four-function calculators of the 1970s to the modern laptops. For example, if each transistor takes up a space of say 10 μm $\times$ 10 μm, then we could fit 9 million of them into a chip of size 3 cm $\times$ 3 cm, since

$$\frac{3 \text{ cm}}{10 \ \mu\text{m}} = 3000 \quad \rightarrow \quad 3000 \times 3000 = 9 \text{ million.}$$

That is where things stood back in the ancient 1990s. But now that a transistor takes up an area of ~ 1 μm $\times$ 1 μm, we can fit 900 million (nearly a billion) of them into the same 3 cm $\times$ 3 cm chip. Where things will go from here remains unclear, since there are major roadblocks to continued miniaturization, the most obvious of which is the difficulty of dissipating the heat that is generated. Any laptop user knows how hot it gets when it is working hard, and it seems difficult to increase the number of switches or their speed too much further.

This book, however, is not about the amazing feats of microelectronics or where the field might be headed. It is about a less-appreciated by-product of the microelectronics revolution, namely the deeper understanding of current flow, energy exchange and device operation that it has enabled, which has inspired the perspective described in this book. Let me explain what we mean.

1.1 Conductance

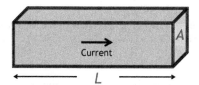

A basic property of a conductor is its resistance R which is related to the cross-sectional area A and the length L by the relation

$$R = \frac{V}{I} = \frac{\rho L}{A} \tag{1.1a}$$

$$G = \frac{I}{V} = \frac{\sigma A}{L}. \tag{1.1b}$$

The resistivity ρ is a geometry-independent property of the material that the channel is made of. The reciprocal of the resistance is the conductance G which is written in terms of the reciprocal of the resistivity called the conductivity σ. So what determines the conductivity?

Our usual understanding is based on the view of electronic motion through a solid as "diffusive" which means that the electron takes a random walk from the source to the drain, traveling in one direction for some length of time before getting scattered into some random direction as sketched in Fig. 1.3. The mean free path, that an electron travels before getting scattered is typically less than a micrometer (also called a micron $= 10^{-3}$ mm, denoted μm) in common semiconductors, but it varies widely with temperature and from one material to another.

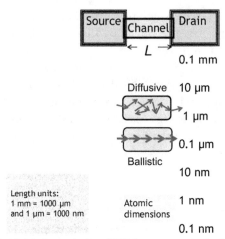

Fig. 1.3   The length of the channel of an FET has progressively shrunk with every new generation of devices ("Moore's law") and stands today at 14 nm, which amounts to ~ 100 atoms.

It seems reasonable to ask what would happen if a resistor is shorter than a mean free path so that an electron travels ballistically ("like a bullet") through the channel. Would the resistance still be proportional to length as described by Eq. (1.1a)? Would it even make sense to talk about its resistance?

These questions have intrigued scientists for a long time, but even twenty five years ago one could only speculate about the answers. Today the answers are quite clear and experimentally well established. Even the transistors in commercial laptops now have channel lengths $L \sim 14$ nm, corresponding to a few hundred atoms in length! And in research laboratories people have even measured the resistance of a hydrogen molecule.

1.2   Ballistic Conductance

It is now clearly established that the resistance R_B and the conductance G_B of a ballistic conductor can be written in the form

$$R_B = \frac{h}{q^2}\frac{1}{M} \simeq 25 \text{ k}\Omega \times \frac{1}{M} \tag{1.2a}$$

$$G_B = \frac{q^2}{h}M \simeq 40 \text{ } \mu\text{S} \times M \tag{1.2b}$$

where q, h are fundamental constants and M represents the number of effective channels available for conduction. Note that we are now using the word "channel" not to denote the physical channel in Fig. 1.3, but in the sense of parallel paths whose meaning will be clarified in the first two parts of this book. In future we will refer to M as the number of "modes", *a concept that is arguably one of the most important lessons of nanoelectronics and mesoscopic physics.*

1.3 What Determines the Resistance?

The ballistic conductance G_B (Eq. (1.2b)) is now fairly well-known, but the common belief is that it is relevant only for short conductors and belongs in a course on special topics like mesoscopic physics or nanoelectronics. We argue that the resistance for both long and short conductors can be written in terms of G_B (λ: mean free path)

$$G = \frac{G_B}{\left(1 + \dfrac{L}{\lambda}\right)}. \tag{1.3}$$

Ballistic and diffusive conductors are not two different worlds, but rather a continuum as the length L is increased. For $L \ll \lambda$, Eq. (1.3) reduces to $G \simeq G_B$, while for $L \gg \lambda$,

$$G \simeq \frac{G_B \lambda}{L},$$

which morphs into Ohm's law (Eq. (1.1b)) if we write the conductivity as

$$\sigma = \frac{GL}{A} = \frac{G_B}{A}\lambda = \frac{q^2}{h}\frac{M}{A}\lambda \qquad (New\ Expression). \tag{1.4}$$

The conductivity of long diffusive conductors is determined by the number of modes per unit area (M/A) which represents a basic material property that is reflected in the conductance of ballistic conductors.

By contrast, the standard expressions for conductivity are all based on bulk material properties. For example freshman physics texts typically describe the Drude formula (momentum relaxation time: τ_m):

$$\sigma = q^2 \frac{n}{m} \tau_m \qquad (Drude\ formula) \tag{1.5}$$

involving the effective mass (m) and the density of free electrons (n). This is the equation that many researchers carry in their head and use to interpret experimental data. However, it is tricky to apply if the electron dynamics is not described by a simple positive effective mass m. A more general but less well-known expression for the conductivity involves the density of states (D) and the diffusion coefficient $(\overline{D})$

$$\sigma = q^2 \frac{D}{AL} \overline{D} \qquad (Degenerate\ Einstein\ relation). \qquad (1.6)$$

In Part A of this book we used fairly elementary arguments to establish the new formula for conductivity given by Eq. (1.4) and show its equivalence to Eq. (1.6). In Part A we also introduced an energy band model and related Eqs. (1.4) and (1.6) to the Drude formula (Eq. (1.5)) under the appropriate conditions when an effective mass can be defined.

We could combine Eqs. (1.3) and (1.4) to say that the standard Ohm's law (Eqs. (1.1)) should be replaced by the result

$$G = \frac{\sigma A}{L + \lambda} \rightarrow R = \frac{\rho}{A}\,(L + \lambda), \qquad (1.7)$$

suggesting that the ballistic resistance (corresponding to $L \ll \lambda$) is equal to $\rho\lambda/A$ which is the resistance of a channel with resistivity ρ and length equal to the mean free path λ.

But this can be confusing since neither resistivity nor mean free path are meaningful for a ballistic channel. It is just that the resistivity of a diffusive channel is inversely proportional to the mean free path, and the product $\rho\lambda$ is a material property that determines the ballistic resistance R_B. A better way to write the resistance is from the inverse of Eq. (1.3):

$$R = R_B \left(1 + \frac{L}{\lambda}\right). \qquad (1.8)$$

This brings us to a key conceptual question that caused much debate and discussion in the 1980s and still seems less than clear! Let me explain.

1.4   Where is the Resistance?

Equation (1.8) tells us that the total resistance has two parts

$$\underbrace{R_B}_{\text{length-independent}} \qquad \text{and} \qquad \underbrace{\frac{R_B L}{\lambda}}_{\text{length-dependent}}.$$

It seems reasonable to assume that the length-dependent part is associated with the channel. What is less clear is that the length-independent part (R_B) is associated with the interfaces between the channel and the two contacts as shown in Fig. 1.4.

How can we split up the overall resistance into different components and pinpoint them spatially? If we were talking about a large everyday resistor, the approach is straightforward: we simply look at the voltage drop across the structure. Since the same current flows everywhere, the voltage drop at any point should be proportional to the resistance at that point $\Delta V = I \Delta R$. A resistance localized at the interface should also give a voltage drop localized at the interface as shown in Fig. 1.4.

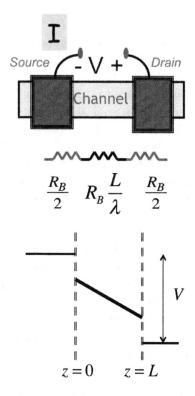

Fig. 1.4   The length-dependent part of the resistance in Eq. (1.8) is associated with the channel while the length-independent part is associated with the interfaces between the channel and the two contacts. Shown below is the spatial profile of the "potential" which supports the spatial distribution of resistances shown.

What makes this discussion not so straightforward in the context of nanoscale conductors is that it is not obvious how to draw a spatial potential profile on a nanometer scale. The key question is well-known in the context of electronic devices, namely the distinction between the electrostatic potential and the electrochemical potential.

The former is related to the electric field F

$$F = -\frac{d\phi}{dz},$$

since the force on an electron is qF, it seems natural to think that the current should be determined by $d\phi/dz$. However, it is well-recognized that this is only of limited validity at best. More generally current is driven by the gradient in the *electrochemical potential*:

$$\frac{I}{A} \equiv J = -\frac{\sigma}{q}\frac{d\mu}{dz}. \qquad (1.9)$$

Just as heat flows from higher to lower temperatures, electrons flow from higher to lower electrochemical potentials giving an electron current that is proportional to $-d\mu/dz$. It is only under special conditions that μ and ϕ track each other and one can be used in place of the other. Although the importance of electrochemical potentials and quasi-Fermi levels is well established in the context of device physics, many experts feel uncomfortable about using these concepts on a nanoscale and prefer to use the electrostatic potential instead. However, I feel that this obscures the underlying physics and considerable conceptual clarity can be achieved by defining electrochemical potentials and quasi-Fermi levels carefully on a nanoscale.

The basic concepts are now well established with careful experimental measurements of the potential drop across nanoscale defects (see for example, Willke *et al.*, 2015). Theoretically it was shown using a full quantum transport formalism (which we discuss in part B) that a suitably defined electrochemical potential shows abrupt drops at the interfaces, while the corresponding electrostatic potential is smoothed out over a screening length making the resulting drop less obvious (Fig. 1.5). These ideas are described in simple semiclassical terms (following Datta, 1995) in Part 3 of this volume.

1.5   But Where is the Heat?

One often associates the electrochemical potential with the energy of the electrons, but at the nanoscale this viewpoint is completely incompatible with what we are discussing. The problem is easy to see if we consider an ideal ballistic channel with a defect or a barrier in the middle, which is the problem Rolf Landauer posed in 1957.

Common sense says that the resistance is caused largely by the barrier and we will show in Chapter 10 that a suitably defined electrochemical

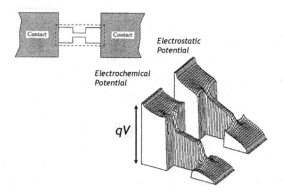

Fig. 1.5 Spatial profile of electrostatic and electrochemical potentials in a nanoscale conductor using a quantum transport formalism. Reproduced from McLennan *et al.*, 1991.

potential indeed shows a spatial profile that shows a sharp drop across the barrier in addition to abrupt drops at the interfaces as shown in Fig. 1.6.

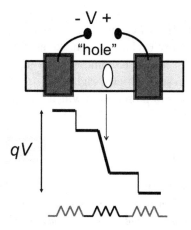

Fig. 1.6 Potential profile across a ballistic channel with a hole in the middle.

If we associate this electrochemical potential with the energy of the electrons then an abrupt potential drop across the barrier would be accompanied by an abrupt drop in the energy, implying that heat is being dissipated locally at the scatterer. This requires the energy to be transferred from the electrons to the lattice so as to set the atoms jiggling which manifests itself as heat. But a scatterer does not necessarily have the degrees of freedom needed to dissipate energy: it could for example be just a hole in the middle of the channel with no atoms to "jiggle".

In short, the resistance R arises from the loss of momentum caused in this case by the "hole" in the middle of the channel. But the dissipation I^2R could occur very far from the hole and the potential in Fig. 1.6 cannot represent the energy. So what does it represent?

The answer is that the electrochemical potential represents the degree of filling of the available states, so that it indicates the number of electrons and not their energy. It is then easy to understand the abrupt drop across a barrier which represents a bottleneck on the electronic highway. As we all know there are traffic jams right before a bottleneck, but as soon as we cross it, the road is all empty: that is exactly what the potential profile in Fig. 1.6 indicates!

In short, everyone would agree that a "hole" in an otherwise ballistic channel is the cause and location of the resulting resistance and an electrochemical potential defined to indicate the number of electrons correlates well with this intuition. But this does not indicate the location of the dissipation I^2R.

The hole in the channel gives rise to "hot" electrons with a non-equilibrium energy distribution which relaxes back to normal through a complex process of energy exchange with the surroundings over an energy relaxation length $L_E \sim$ tens of nanometers or longer. The process of dissipation may be of interest in its own right, but it does not help locate the hole that caused the loss of momentum which gave rise to resistance in the first place.

1.6 Elastic Resistors

Once we recognize the spatially distributed nature of dissipative processes it seems natural to model nanoscale resistors shorter than L_E as an *ideal elastic resistor which we define as one in which all the energy exchange and dissipation occurs in the contacts and none within the channel itself* (Fig. 1.7).

For a ballistic resistor R_B, as my colleague Ashraf often points out, it is almost obvious that the corresponding Joule heat I^2R must occur in the contacts. After all a bullet dissipates most of its energy to the object it hits rather than to the medium it flies through.

There is experimental evidence that real nanoscale conductors do actually come close to this idealized model which has become widely used

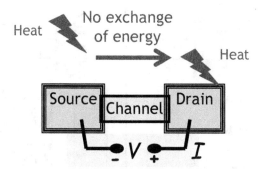

Fig. 1.7 *The ideal elastic resistor* with the Joule heat $VI = I^2R$ generated entirely in the contacts as sketched. Many nanoscale conductors are believed to be close to this ideal.

ever since the advent of mesoscopic physics in the late 1980s and is often referred to as the *Landauer approach*. However, it is generally believed that this viewpoint applies only to near-ballistic transport and to avoid this association we are calling it an *elastic resistor* rather than a *Landauer resistor*.

What we wish to stress is that even a diffusive conductor full of "potholes" that destroy momentum could in principle dissipate all the Joule heat in the contacts. And even if it does not, its resistance can be calculated accurately from an idealized model that assumes it does. Indeed we will use this elastic resistor model to obtain the conductivity expression in Eq. (1.4) and show that it agrees well with the standard results.

But surely we cannot ignore all the dissipation inside a long resistor and calculate its resistance accurately treating it as an elastic resistor? We believe we can do so in many cases of interest, especially at low bias. The underlying issues can be understood qualitatively using the simple circuit model shown in Fig. 1.8. For an elastic resistor each energy channel E_1, E_2 and E_3 is independent with no flow of electrons between them as shown on the left. Inelastic processes induce "vertical" flow between the energy channels represented by the vertical resistors as shown on the right. When can we ignore the vertical resistors?

If the series of resistors representing individual channels are identical, then the nodes connected by the vertical resistors will be at the same potential, so that *there will be no current flow through them*. Under these conditions, an elastic resistor model that ignores the vertical resistors is quite accurate.

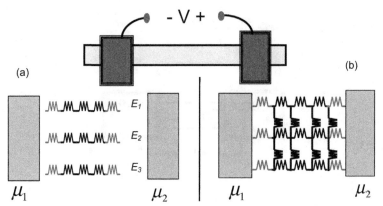

Fig. 1.8   A simple circuit model: (a) For elastic resistors, individual energy channels E_1, E_2 and E_3 are decoupled with no flow between them. (b) Inelastic processes cause vertical flow between energy channels through the additional resistors shown.

But vertical flow cannot always be ignored. For example, Fig. 1.9a shows a conductor where the lower energy levels E_2 and E_3 conduct poorly compared to E_1. We would then expect the electrons to flow upwards in energy on the left and downwards in energy on the right as shown, thus cooling the lattice on the left and heating the lattice on the right, leading to the well-known *Peltier effect* discussed in Chapter 13.

The role of vertical flow can be even more striking if the left contact connects only to the channel E_1 while the right contact connects only to E_3 as shown in Fig. 1.9b. No current can flow in such a structure without vertical flow, and the entire current is purely a vertical current. This is roughly what happens in p-n junctions which is discussed a little further in Section 12.1.

The bottom line is that elastic resistors generally provide a good description of short conductors and the Landauer approach has become quite common in mesoscopic physics and nanoelectronics. What is not well recognized is that this approach can provide useful results even for long conductors. In many cases, but not always, we can ignore inelastic processes and calculate the resistance quite accurately as long as the momentum relaxation has been correctly accounted for, as discussed further in Section 3.3.

But why would we want to ignore inelastic processes? Why is the theory of elastic resistors any more straightforward than the standard approach? To understand this we first need to talk briefly about the transport theories on which the standard approach is based.

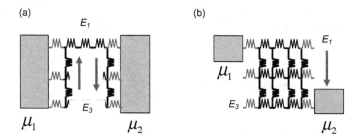

Fig. 1.9   Two examples of structures where vertical flow between energy channels can be important: (a) If the lower energy levels E_2 and E_3 conduct poorly, electrons will flow up in energy on the left and down in energy on the right as shown. (b) If the left contact couples to an upper energy E_1 while the right contact couples to a lower energy E_3, then the current flow is purely vertical, occurring *only* through inelastic processes.

1.7   Transport Theories

Flow or transport always involves two fundamentally different types of processes, namely elastic transfer and heat generation, belonging to two distinct branches of physics. The first involves frictionless mechanics of the type described by Newton's laws or the Schrödinger equation. The second involves the generation of heat described by the laws of thermodynamics.

The first is driven by forces or potentials and is reversible. The second is driven by entropy and is irreversible. Viewed in reverse, entropy-driven processes look absurd, like heat flowing spontaneously from a cold to a hot surface or an electron accelerating spontaneously by absorbing heat from its surroundings.

Normally the two processes are intertwined and a proper description of current flow in electronic devices requires the advanced methods of non-equilibrium statistical mechanics that integrate mechanics with thermodynamics. Over a century ago Boltzmann taught us how to combine Newtonian mechanics with heat generating or entropy-driven processes and the resulting Boltzmann transport equation (BTE) is widely accepted as the cornerstone of semiclassical transport theory. The word semiclassical is used because some quantum effects have also been incorporated approximately into the same framework.

$$\text{Classical Dynamics} \quad + \quad \text{⚡} \quad = \quad \text{BTE}$$

A full treatment of quantum transport requires a formal integration of quantum dynamics described by the Schrödinger equation with heat generating processes.

This is exactly what is achieved in the non-equilibrium Green's function (NEGF) method originating in the 1960s from the seminal works of Martin and Schwinger (1959), Kadanoff and Baym (1962), Keldysh (1965) and others.

1.7.1 *Why elastic resistors are conceptually simpler*

The BTE takes many semesters to master and the full NEGF formalism, even longer. Much of this complexity comes from the subtleties of combining mechanics with distributed heat-generating processes.

The operation of the elastic resistor can be understood in far more elementary terms because of the clean spatial separation between the force-driven and the entropy-driven processes. The former is confined to the channel and the latter to the contacts. As we will see in the next few chapters, the latter is easily taken care of, indeed so easily that it is easy to miss the profound nature of what is being accomplished.

Even quantum transport can be discussed in relatively elementary terms using this viewpoint. For example, Fig. 1.10 shows a plot of the spatial profile of the electrochemical potential across our structure from Fig. 1.6 with a hole in the middle, calculated both from the semiclassical BTE (Chapter 9) and from the NEGF method (part B).

For the NEGF method we show three options. First a coherent model (left) that ignores all interaction within the channel showing oscillations indicative of standing waves. Once we include phase relaxation, the constructive and destructive interferences are lost and we obtain the result in the middle which approaches the semiclassical result. If the interactions

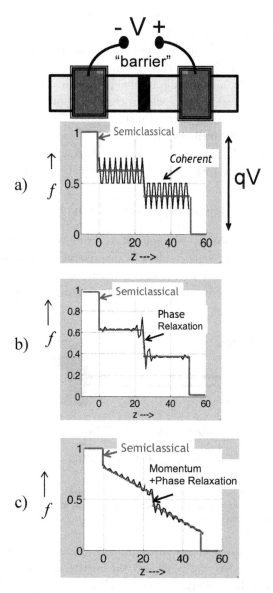

Fig. 1.10 Spatial profile of the electrochemical potential across a channel with a barrier. Solid red line indicates semiclassical result from BTE (part A). Also shown are the results from NEGF (part B) assuming (a) coherent transport, (b) transport with phase relaxation, (c) transport with phase and momentum relaxation. Note that no energy relaxation is included in any of these calculations.

include momentum relaxation as well we obtain a profile indicative of an additional distributed resistance.

None of these models includes energy relaxation and they all qualify as elastic resistors making the theory much simpler than a full quantum transport model that includes dissipative processes. Nevertheless, they all exhibit a spatial variation in the electrochemical potential consistent with our intuitive understanding of resistance.

A good part of my own research in the past was focused in this area developing the NEGF method, but we will get to it only in part B after we have "set the stage" in this volume using a semiclassical picture.

1.8   Is Transport Essentially a Many-body Process?

The idea that resistance can be understood from a model that ignores interactions within the channel comes as a surprise to many, possibly because of an interesting fact that we all know: when we turn on a switch and a bulb lights up, it is not because individual electrons flow from the switch to the bulb. That would take far too long.

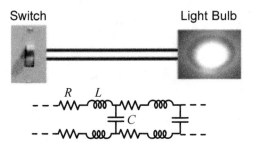

Fig. 1.11   To describe the propagation of signals we need a distributed RLC, model that includes an inductance L and a capacitance C which are ordinarily determined by magnetostatics and electrostatics respectively.

The actual process is nearly instantaneous because one electron pushes the next, which pushes the next and the disturbance travels essentially at the speed of light. Surely, our model that localizes all interactions at arbitrarily placed contacts (Fig. 3.5 of Part A, see Appendix F) cannot describe this process?

The answer is that to describe the propagation of transient signals we need a model that includes not just a resistance R, but also an inductance L and a capacitance C as shown in Fig. 1.11. These could include transport related corrections in small conductors but are ordinarily determined by magnetostatics and electrostatics respectively (Salahuddin *et al.*, 2005).

In this distributed *RLC* transmission line, the signal velocity determined by L and C can be well in excess of individual electron velocities reflecting a collective process. However, L and C play no role at low frequencies, since the inductor is then like a "short circuit" and the capacitor is like an "open circuit". The low frequency conduction properties are represented solely by the resistance R and can usually be understood fairly well in terms of the transport of individual electrons along M parallel modes (see Eqs. (1.2)) or "channels", a concept that has emerged from decades of research. To quote Phil Anderson from a volume commemorating 50 years of Anderson localization (see Anderson (2010)):

" ... *What might be of modern interest is the "channel" concept which is so important in localization theory. The transport properties at low frequencies can be reduced to a sum over one-dimensional "channels" ...* "

Even though high frequency signals propagate at the "speed of light", there can be no steady-state flow of charge unless an electron transmits from one end to the other, or as Landauer put it, *conductance is transmission*. However, this observation about steady-state currents applies only to *charge* and not to other quantities like *spin*.

1.9 A Different Physical Picture

Let me conclude this overview with an obvious question: why should we bother with idealized models and approximate physical pictures? Can't we simply use the BTE and the NEGF equations which provide rigorous frameworks for describing semiclassical and quantum transport respectively? The answer is yes, and all the results we discuss are benchmarked against the BTE and the NEGF.

However, as Feynman (1963) noted in his classic lectures, even when we have an exact mathematical formulation, we need an intuitive physical picture:

".. *people* .. *say* .. *there is nothing which is not contained in the equations* .. *if I understand them mathematically inside out, I will understand the physics inside out. Only it doesn't work that way.* .. *A physical understanding is a completely unmathematical, imprecise and inexact thing, but absolutely necessary for a physicist.*"

Indeed, most researchers carry a physical picture in their head and it is usually based on the Drude formula (Eq. (1.5)). In this book we will show that an alternative picture based on elastic resistors leads to a formula (Eq. (1.4)) that is more generally valid.

Unlike the Drude formula which treats the electric field as the driving term, this new approach more correctly treats the electrochemical potential as the driving term. This is well-known at the macroscopic level, but somehow seems to have been lost in nanoscale transport, where people cite the difficulty of defining electrochemical potentials. However, that does not justify using electric field as a driving term, an approach that *does not work for inhomogeneous conductors on any scale*.

Since all conductors are fundamentally inhomogeneous on an atomic scale it seems questionable to use electric field as a driving term. We argue that at least for low bias transport, it is possible to define electrochemical potentials or quasi-Fermi levels on an atomic scale and this can lend useful insight into the physics of current flow and the origin of resistance. We believe this is particularly timely because future electronic devices will require a clear understanding of the different potentials.

For example, recent work on spintronics has clearly established experimental situations where upspin and downspin electrons have different electrochemical potentials (sometimes called quasi-Fermi levels) and could even flow in opposite directions because their $d\mu/dz$ have opposite signs. This cannot be understood if we believe that currents are driven by electric fields, $-d\phi/dz$, since up and down spins both see the same electric field and have the same charge. We can expect to see more and more such examples that use novel contacts to manipulate the quasi-Fermi levels of different group of electrons (see Chapter 12 of Part A for further discussion).

In short we believe that the lessons of nanoelectronics lead naturally to a new viewpoint, one that changes even some basic concepts we all learn in freshman physics. This viewpoint represents a departure from the established mindset and I hope it will provide a complementary perspective to facilitate the insights needed to take us to the next level of discovery and innovation.

PART 1

Contact-ing Schrödinger

Chapter 17

The Model

Related video lectures available at course website, Unit 1: L1.1 and Unit 1: L1.10.

Over a century ago Boltzmann taught us how to combine Newtonian mechanics with entropy-driven processes and the resulting Boltzmann trans-

$$\text{Classical Dynamics} \quad + \quad \text{⚡} \quad = \quad \text{BTE}$$

port equation (BTE) is widely accepted as the cornerstone of semiclassical transport theory. Most of the results we have discussed so far can be (and generally are) obtained from the Boltzmann equation, but the concept of an elastic resistor makes them more transparent by spatially separating force-driven processes in the channel from the entropy-driven processes in the contacts.

In this part of this book I would like to discuss the quantum version of this problem, using the non-equilibrium Green's function (NEGF) method to combine quantum mechanics described by the Schrödinger equation with "contacts" much as Boltzmann taught us how to combine classical dynamics with "contacts".

$$\text{Quantum Dynamics} \quad + \quad \text{⚡} \quad = \quad \text{NEGF}$$

The NEGF method originated from the classic works in the 1960s that used the methods of many-body perturbation theory to describe the distributed entropy-driven processes along the channel. Like most of the work

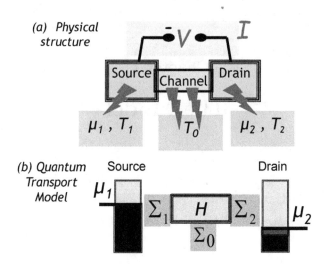

Fig. 17.1　(a) Generic device structure that we have been discussing. (b) General quantum transport model with elastic channel described by a Hamiltonian **H** and its connection to each "contact" described by a corresponding self-energy **Σ**.

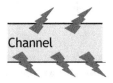

on transport theory (semiclassical or quantum) prior to the 1990s, it was a *"contact-less"* approach focused on the interactions occurring throughout the channel, in keeping with the general view that the physics of resistance lay essentially in these distributed entropy generating processes.

As with semiclassical transport, our discussion starts at the other end with the elastic resistor with entropy-driven processes confined to the contacts. This makes the theory less about interactions and more about *"connecting contacts to the Schrödinger equation"*, or more simply, about *contact-ing Schrödinger*.

But let me put off talking about the NEGF model till the next chapter, and use subsequent chapters to illustrate its application to interesting problems in quantum transport. As indicated in Fig. 17.1b the NEGF method requires two types of inputs: the Hamiltonian, **H** describing the dynamics of an elastic channel, and the self-energy **Σ** describing the connection to the contacts, using the word "contacts" in a broad figurative sense to denote all kinds of entropy-driven processes. Some of these contacts are physical like the ones labeled "1" and "2" in Fig. 17.1b, while some are conceptual like the one labeled "0" representing entropy changing processes distributed throughout the channel.

In this chapter let me just try to provide a super-brief but self-contained introduction to how one writes down the Hamiltonian $\mathbf{H}$. The $\mathbf{\Sigma}$ can be obtained by imposing the appropriate boundary conditions and will be described in later chapters when we look at specific examples applying the NEGF method.

We will try to describe the procedure for writing down $\mathbf{H}$ so that it is accessible even to those who have not had the benefit of a traditional multi-semester introduction to quantum mechanics. Moreover, our emphasis here is on something that may be helpful even for those who have this formal background. Let me explain.

Most people think of the Schrödinger equation as a differential equation which is the form we see in most textbooks. However, practical calculations are usually based on a discretized version that represents the differential equation as a matrix equation involving the Hamiltonian matrix $\mathbf{H}$ of size $N \times N$, N being the number of "*basis functions*" used to represent the structure.

This matrix $\mathbf{H}$ can be obtained from first principles, but a widely used approach is to represent it in terms of a few parameters which are chosen to match key experiments. Such semi-empirical approaches are often used because of their convenience and because they can often explain a wide range of experiments beyond the key ones that are used as input, suggesting that they capture a lot of essential physics.

In order to follow the rest of the chapter it is important for the readers to get a feeling for how one writes down this matrix $\mathbf{H}$ given an accepted energy-momentum $E(\mathbf{p})$ relation (Chapter 6) for the material that is believed to describe the dynamics of conduction electrons with energies around the electrochemical potential .

But I should stress that the NEGF framework we will talk about in subsequent chapters goes far beyond any specific model that we may choose to use for $\mathbf{H}$. The same equations could be (and have been) used to describe say conduction through molecular conductors using first principles Hamiltonians.

17.1   Schrödinger Equation

Related video lecture available at course website, Unit 1: L1.2.

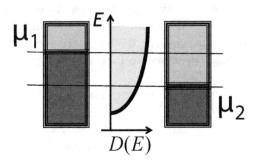

We started this book by noting that the key input needed to understand current flow is the density of states, $D(E)$ telling us the number of states available for an electron to access on its way from the source to the drain.

Theoretical models for $D(E)$ all start from the Schrödinger equation which tells us the available energy levels. However, we managed to obtain expressions for $D(E)$ in Chapter 6 without any serious brush with quantum mechanics by **(1)** starting from a given energy-momentum relation $E(p)$, **(2)** relating the momentum to the wavelength through the *de Broglie relation* $(p \doteq h/\text{wavelength})$ and then **(3)** requiring an integer number of half wavelengths to fit into the conductor, the same way acoustic waves fit on a guitar string.

This heuristic principle is mathematically implemented by writing a wave equation which is obtained from a desired energy-momentum relation by making the replacements

$$E \rightarrow i\hbar \frac{\partial}{\partial t}, \quad \mathbf{p} \rightarrow -i\hbar \nabla \qquad (17.1)$$

where the latter stands for

$$p_x \rightarrow -i\hbar \frac{\partial}{\partial x}, \quad p_y \rightarrow -i\hbar \frac{\partial}{\partial y}, \quad p_z \rightarrow -i\hbar \frac{\partial}{\partial z}.$$

Using this principle, the classical energy-momentum relation

$$E_{\text{classical}}(\mathbf{p}) = \frac{p_x^2 + p_y^2 + p_z^2}{2m} \qquad (17.2a)$$

leads to the wave equation

$$i\hbar \frac{\partial}{\partial t}\tilde{\psi}(x,y,z,t) = -\frac{\hbar^2}{2m}\left(\frac{\partial^2}{\partial x^2} + \frac{\partial^2}{\partial y^2} + \frac{\partial^2}{\partial z^2}\right)\tilde{\psi}(x,y,z,t) \qquad (17.2b)$$

whose solutions can be written in the form of exponentials of the form

$$\tilde{\psi}(x, y, z, t) = \psi_0 \, e^{+ik_x x} \, e^{+ik_y y} \, e^{+ik_z z} \, e^{-iEt/\hbar} \qquad (17.3)$$

where the energy E is related to the wavevector $\mathbf{k}$ by the dispersion relation

$$E\left(\mathbf{k}\right) = \frac{\hbar^2 (k_x^2 + k_y^2 + k_z^2)}{2m} \qquad (17.4)$$

Eq. (17.4) looks just like the classical energy-momentum relation (Eq. (17.2a)) of the corresponding particle with

$$\mathbf{p} = \hbar \mathbf{k} \qquad (17.5)$$

which relates the particulate property $\mathbf{p}$ with the wavelike property $\mathbf{k}$. This can be seen to be equivalent to the de Broglie relation ($p = h/\text{wavelength}$) noting that the wavenumber k is related to the wavelength through

$$k = \frac{2\pi}{wavelength}.$$

The principle embodied in Eq. (17.1) ensures that the resulting wave equation has a group velocity that is the same as the velocity of the corresponding particle

$$\underbrace{\frac{1}{\hbar} \nabla_k E}_{\text{Wave group velocity}} = \underbrace{\nabla_p E}_{\text{Particle velocity}}$$

17.1.1 *Spatially varying potential*

The wave equation Eq. (17.2b) obtained from the energy-momentum relation describes free electrons. If there is a force described by a potential energy $U(\mathbf{r})$ so that the classical energy is given by

$$E_{\text{classical}}\left(\mathbf{r}, \mathbf{p}\right) = \frac{p_x^2 + p_y^2 + p_z^2}{2m} + U(x, y, z) \qquad (17.6a)$$

then the corresponding wave equation has an extra term due to $U(\mathbf{r})$

$$i\hbar \frac{\partial}{\partial t} \tilde{\psi} = -\frac{\hbar^2}{2m} \nabla^2 \tilde{\psi} + U(\mathbf{r}) \tilde{\psi} \qquad (17.6b)$$

where $\mathbf{r} \equiv (x, y, z)$ and the *Laplacian operator* is defined as

$$\nabla^2 \equiv \frac{\partial^2}{\partial x^2} + \frac{\partial^2}{\partial y^2} + \frac{\partial^2}{\partial z^2}.$$

Solutions to Eq. (17.6b) can be written in the form

$$\tilde{\psi}(\mathbf{r}, t) = \psi(\mathbf{r}) \, e^{-iEt/\hbar}$$

where $\psi(\mathbf{r})$ obeys the time-independent Schrödinger equation

$$E\psi(\mathbf{r}) = \mathbf{H}_{op}\psi(\mathbf{r}) \qquad (17.7a)$$

where $\mathbf{H}_{op}$ is a *differential operator* obtained from the classical energy function in Eq. (17.6a), using the replacement mentioned earlier (Eq. (17.1)):

$$\mathbf{H}_{op} = -\frac{\hbar^2}{2m}\nabla^2 + U(\mathbf{r}). \qquad (17.7b)$$

Quantum mechanics started in the early twentieth century with an effort to "understand" the energy levels of the hydrogen atom deduced from the experimentally observed spectrum of the light emitted from an incandescent source. For a hydrogen atom Schrödinger used the potential energy

$$U(\mathbf{r}) = -Z\frac{q^2}{4\pi\,\varepsilon_0 r}$$

where the atomic number $Z = 1$, due to a point nucleus with charge $+q$, and solved Eqs. (17.7) analytically for the allowed energy values E_n (called the eigenvalues of the operator $\mathbf{H}_{op}$) given by

$$E_n = -\frac{Z^2}{n^2}\frac{q^2}{8\pi\,\varepsilon_0 a_0} \qquad (17.8)$$

with

$$a_0 = \frac{4\pi\,\varepsilon_0\hbar^2}{m\,q^2}$$

and the corresponding solutions

$$\psi_{n\ell m}(\mathbf{r}) = R_{n\ell}(\mathbf{r})\,Y_\ell^m(\theta,\phi)$$

obeying the equation

$$E_n\,\psi_{n\ell m}(\mathbf{r}) = \left(-\frac{\hbar^2}{2m}\nabla^2 - \frac{Zq^2}{4\pi\varepsilon_0 r}\right)\psi_{n\ell m}(\mathbf{r}).$$

The energy eigenvalues in Eq. (17.8) were in extremely good agreement with the known experimental results, leading to general acceptance of the Schrödinger equation as **the** wave equation describing electrons, just as acoustic waves, for example, on a guitar string are described by

$$\omega^2 u(z) = -\frac{\partial^2}{\partial z^2}u.$$

A key point of similarity to note is that when a guitar string is clamped between two points, it is able to vibrate only at discrete frequencies determined by the length L. Similarly electron waves when "clamped" have

$$2p$$

$$n = 2, \ell = 1, m = -1, 0, +1$$

$$n = 2, \ell = 0, m = 0$$

$$2s$$

$$n = 1, \ell = 0, m = 0$$

$$1s$$

Fig. 17.2 Energy levels in atoms are catalogued with three indices n, l, and m.

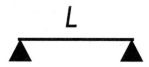

discrete energies and most quantum mechanics texts start by discussing the corresponding "particle in a box" problem.

Shorter the length L, higher the pitch of a guitar and hence the spacing between the harmonics. Similarly smaller the box, greater the spacing between the allowed energies of an electron. Indeed one could view the hydrogen atom as an extremely small 3D box for the electrons giving rise to the discrete energy levels shown in Fig. 17.2. This is of course just a qualitative picture. Quantitatively, we have to solve the time-independent Schrödinger equation (Eq. (17.7)).

There is also a key dissimilarity between classical waves and electron waves. For acoustic waves we all know what the quantity $u(z)$ stands for: it is the displacement of the string at the point z, something that can be readily measured. By contrast, the equivalent quantity for electrons, $\psi(\mathbf{r})$ (called its wavefunction), is a complex quantity that cannot be measured directly and it took years for scientists to agree on its proper interpretation. The present understanding is that the real quantity $\psi\psi^*$ describes the *probability* of finding an electron in a unit volume around $\mathbf{r}$. This quantity, when summed over many electrons, can be interpreted as the *average electron density.*

17.2   Electron-electron Interactions and the SCF Method

After the initial success of the Schrödinger equation in "explaining" the experimentally observed energy levels of the Hydrogen atom, scientists applied it to increasingly more complicated atoms and by 1960 had achieved good agreement with experimentally measured results for all atoms in the periodic table (Herman and Skillman (1963)). It should be noted, however, that these calculations are far more complicated primarily because of the need to include the electron-electron (e-e) interactions in evaluating the potential energy (Hydrogen has only one electron and hence no e-e interactions).

For example, Eq. (17.8) gives the lowest energy for a Hydrogen atom as $E_1 = -13.6$ eV in excellent agreement with experiment. It takes a photon with at least that energy to knock the electron out of the atom ($E > 0$), that is to cause photoemission. Looking at Eq. (17.8) one might think that in Helium with $Z = 2$, it would take a photon with energy $\sim 4 \times 13.6$ eV $=$ 54.5 eV to knock an electron out. However, it takes photons with far less energy ~ 30 eV and the reason is that the electron is repelled by the other electron in Helium. However, if we were to try to knock the second electron out of Helium, it would indeed take photons with energy ~ 54 eV, which is known as the second ionization potential. But usually what we want is the first ionization potential or a related quantity called the electron affinity. Let me explain.

Current flow involves adding an electron from the source to the channel and removing it into the drain. However, these two events could occur in either order.

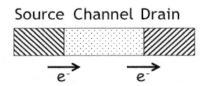

The electron could first be added and then removed so that the channel evolves as follows

A. $N \rightarrow N+1 \rightarrow N$ electrons (Affinity levels).

But if the electron is first removed and then added, the channel would evolve as

B. $N \rightarrow N-1 \rightarrow N$ electrons (Ionization levels).

In the first case, the added electron would feel the repulsive potential due to N electrons. Later when removing it, it would still feel the potential due to N electrons since no electron feels a potential due to itself. So the electron energy levels relevant to this process should be calculated from the Schrödinger equation using a repulsive potential due to N electrons. These are known as the *affinity levels*.

In the second case, the removed electron would feel the repulsive potential due to the other $N - 1$ electrons. Later when adding an electron, it would also feel the potential due to $N - 1$ electrons. So the electron energy levels relevant to this process should be calculated from the Schrödinger equation using a repulsive potential due to $N - 1$ electrons. These are known as the *ionization levels*.

The difference between the two sets of levels is basically the difference in potential energy due to one electron, called the *single electron charging energy* U_0. For something as small as a Helium atom it is ~ 25 eV, so large that it is hard to miss. For large conductors it is often so small that it can be ignored, and it does not matter too much whether we use the potential due to N electrons or due to $N - 1$ electrons. For small conductors, under certain conditions the difference can be important giving rise to single-electron charging effects, which we will ignore for the moment and take up again later in Chapter 22.

Virtually all the progress that has been made in understanding "condensed matter," has been based on the self-consistent field (SCF) method where we think of each electron as behaving quasi-independently feeling an average self-consistent potential $U(\mathbf{r})$ due to all the other electrons in addition to the nuclear potential. This potential depends on the electron density $n(\mathbf{r})$ which in turn is determined by the wavefunctions of the filled states. Given the electron density how one determines $U(\mathbf{r})$ is the subject of much discussion and research. The "zero order" approach is to calculate $U(\mathbf{r})$ from $n(\mathbf{r})$ based on the laws of electrostatics, but it is well-established that this so-called *Hartree approximation* will overestimate the repulsive potential and there are various approaches for estimating this reduction. The *density functional theory* (DFT) has been spectacularly successful in describing this correction for equilibrium problems and in its simplest form amounts to a reduction by an amount proportional to the cube root of the electron density

$$U(\mathbf{r}) = U_{\text{Hartree}} - \frac{q^2}{4\pi\varepsilon} \left(n\left(\mathbf{r}\right) \right)^{1/3}. \qquad (17.9)$$

Many are now using similar corrections for non-equilibrium problems like current flow as well, though we believe there are important issues that remain to be resolved.

We should also note that there is a vast literature (both experiment and theory) on a regime of transport that cannot be easily described within an SCF model. It is not just a matter of correctly evaluating the self-consistent potential. The very picture of quasi-independent electrons moving in a self-consistent field needs revisiting, as we will see in Chapter 22.

17.3　Differential to Matrix Equation

Related video lecture available at course website, Unit 1: L1.3.

All numerical calculations typically proceed by turning the differential equation in Eq. (17.7) into a matrix equation of the form

$$E\,\mathbf{S}\psi = \mathbf{H}\psi \tag{17.10a}$$

or equivalently

$$E\sum_m S_{nm}\,\psi_m = \sum_m H_{nm}\,\psi_m \tag{17.10b}$$

by expanding the wavefunction in terms of a set of known functions $u_m(\mathbf{r})$ called the **basis functions:**

$$\psi(\mathbf{r}) = \sum_m \psi_m\,u_m(\mathbf{r}). \tag{17.11a}$$

The elements of the two matrices $\mathbf{S}$ and $\mathbf{H}$ are given respectively by

$$S_{nm} = \int d\mathbf{r}\; u_n^*(\mathbf{r})\,u_m(\mathbf{r}) \tag{17.11b}$$

$$H_{nm} = \int d\mathbf{r}\; u_n^*(\mathbf{r})\,\mathbf{H}_{op}\,u_m(\mathbf{r}). \tag{17.11c}$$

These expressions are of course by no means obvious, but we will not go into it further since we will not really be making any use of them. Let me explain why.

17.3.1 *Semi-empirical tight-binding (TB) models*

There are a wide variety of techniques in use which differ in the specific basis functions they use to convert the differential equation into a matrix equation. But once the matrices $\mathbf{S}$ and $\mathbf{H}$ have been evaluated, the eigenvalues E of Eq. (17.10) (which are the allowed energy levels) are determined using powerful matrix techniques that are widely available. In modeling nanoscale structures, it is common to use basis functions that are spatially localized rather than extended functions like sines or cosines. For example, if we were to model a Hydrogen molecule, with two positive nuclei as shown (see Fig. 17.3), we could use two basis functions, one localized around the left nucleus and one around the right nucleus. One could then work through the algebra to obtain $\mathbf{H}$ and $\mathbf{S}$ matrices of the form

$$\mathbf{H} = \begin{bmatrix} \varepsilon & t \\ t & \varepsilon \end{bmatrix} \quad \text{and} \quad \mathbf{S} = \begin{bmatrix} 1 & s \\ s & 1 \end{bmatrix} \tag{17.12}$$

where ε, t and s are three numbers.

The two eigenvalues from Eq. (17.10) can be written down analytically as

$$E_1 = \frac{\varepsilon - t}{1 - s} \quad \text{and} \quad E_2 = \frac{\varepsilon + t}{1 + s}$$

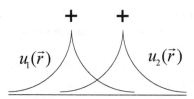

Fig. 17.3 To model a Hydrogen molecule with two positive nuclei, one could use two basis functions, one localized around the left nucleus and one around the right nucleus.

What we just described above would be called a first-principles approach. Alternatively one could adopt a semi-empirical approach treating ε, t and s as three numbers to be adjusted to give the best fit to our "favorite" experiments. For example, if the energy levels $E_{1,2}$ are known from experiments, then we could try to choose numbers that match these. Indeed, it is common to assume that the $\mathbf{S}$ matrix is just an identity matrix ($s = 0$), so that there are only two parameters ε and t which are then adjusted to match $E_{1,2}$. Basis functions with $s = 0$ are said to be "*orthogonal*".

17.3.2   *Size of matrix,* $N = n \times b$

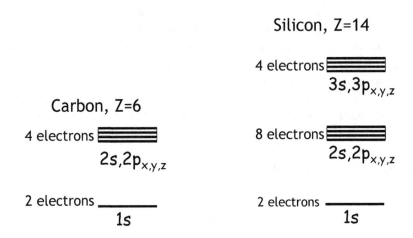

What is the size of the **H** matrix? Answer: $(N \times N)$, N being the total number of basis functions. How many basis functions? Answer: Depends on the approach one chooses. In the tight-binding (TB) approach, which we will use, the basis functions are the atomic wavefunctions for individual atoms, so that $N = n \times b$, n being the number of atoms and b, the number of basis functions per atom. What is b? Let us look at specific examples.

Suppose we want to model current flow through **graphene** consisting of carbon atoms arranged in a two dimensional hexagonal sheet (see Fig. 17.4). Carbon ($Z = 6$) has six electrons which are accommodated in the 1s, 2s and 2p levels as shown. The electrons in the highest levels that is the 2s and 2p levels are the so called valence electrons that move around and carry current. So in the simplest theories, it is common to use the 2s and 2p levels on each atom as the basis functions, with $b = 4$.

The same is true of say silicon ($Z = 14$), the most common semiconductor for electronic devices. Its fourteen electrons are accommodated as shown with the valence electrons in the 3s, 3p levels. Once again in the simplest models $b = 4$, though some models include five 3d levels and/or the two 4s levels as part of the basis functions too.

One of the nice things about graphene is that the 2s, $2p_x$, $2p_y$ orbitals are in the simplest approximation completely decoupled from the $2p_z$ orbitals, and for understanding current flow, one can get a reasonable description with just one $2p_z$ orbital for every carbon atom, so that $b = 1$.

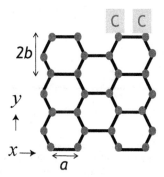

Fig. 17.4 Graphene consists of a two-dimensional sheet of carbon atoms arranged in a two-dimensional hexagonal lattice.

In these simplest models, the matrix **H** is of size $(n \times n)$, n being the total number of carbon atoms. Its diagonal elements have some value ε, while the matrix element H_{nm} equals some value t if n and m happen to be nearest neighbors. If they are not nearest neighbors then one expects the value to be smaller since the functions u_m and u_n appearing in Eqs. (17.11b) and (17.11c) do not overlap as much. In nearest neighbor tight-binding models it is common to set all such matrix elements to zero, so that we are finally left with just two parameters ε and t which are then adjusted to match known results.

17.4 Choosing Matrix Parametrs

One common way to select the parameters is to fit the known energy dispersion relation $E(\mathbf{k})$, also called the energy-momentum relation $E(\mathbf{p})$ (Note that $\mathbf{p} = \hbar \mathbf{k}$) as discussed in Chapter 6. These relations have been arrived at through years of work combining careful experimental measurements with sophisticated first-principles calculations. If we can get our semi-empirical model to fit the accepted dispersion relation for a material, we have in effect matched the whole set of experiments that contributed to it.

17.4.1 *One-dimensional conductor*

Related video lecture available at course website, Unit 1: L1.4.

Suppose we have a one-dimensional conductor that we would like to model with a nearest neighbor orthogonal tight-binding model with two parameters ε and t representing the diagonal elements and the nearest neighbor coupling (Fig. 17.5).

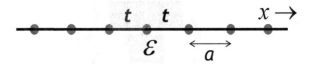

Fig. 17.5   A one-dimensional array of atoms spaced by "a" modeled with a nearest neighbor orthogonal tight-binding model with two parameters ε and t representing the diagonal elements and the nearest neighbor coupling.

How would we choose ε and t so that we approximate a parabolic dispersion relation

$$E(k) = E_c + \frac{\hbar^2 k^2}{2m} \; ? \tag{17.13}$$

The answer is that our model represents a set of algebraic equations (see Eqs. (17.10)) which for the orthogonal model reduces to

$$E\psi_n = \sum_m H_{nm}\psi_m \quad \Rightarrow \quad E = \sum_m H_{nm}\frac{\psi_m}{\psi_n}.$$

If we assume a solution of the form

$$\psi_n = \psi_0 e^{ik\,na}$$

we obtain the $E(k)$ relation corresponding to Eqs. (17.10):

$$E(k) = \sum_m H_{nm} e^{ik(m-n)a}. \tag{17.14}$$

Can we always assume a solution of this form?   No.   In general Eq. (17.14) will give us different results for $E(k)$ depending on what value we choose for n when doing the summation and what we get for some particular choice of n is not very helpful. But if the structure is "*translationally invariant*" such that we get the same answer for all n then we get a unique $E(k)$ relation and $\psi_n = \psi_0 \exp(ikna)$ indeed represents an acceptable solution to our set of equations.

For our particular nearest neighbor model Eq. (17.14) yields straightforwardly

$$E(k) = \varepsilon + t\,e^{+ika} + t\,e^{-ika} = \varepsilon + 2t\,\cos(ka). \tag{17.15}$$

How would we make this match the desired parabolic relation in Eq. (17.13)? Clearly one could not match them for all values of k, only for a limited range. For example, if we want them to match over a range of k-values around $k = 0$, we can expand the cosine in a Taylor series around $ka = 0$ to write

$$\cos(ka) \approx 1 - \frac{(ka)^2}{2}$$

so that the best match is obtained by choosing

$$t = -\frac{\hbar^2}{2ma^2} \tag{17.16a}$$

and

$$\varepsilon = E_c - 2t. \tag{17.16b}$$

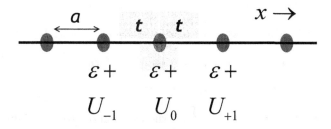

Fig. 17.6   A spatially varying potential $U(x)$ along the channel is included by adding the local value of U to the diagonal element ε.

Finally, I should mention that when modeling a device there could be a spatially varying potential $U(x)$ along the channel which is included by adding the local value of U to the diagonal element as indicated in Fig. 17.6. We now no longer have the *"translational invariance"* needed for a solution of the form $\exp(ikx)$ and the concept of a dispersion relation $E(k)$ is not valid. But a Hamiltonian of the form just described (Fig. 17.6) can be used for numerical calculations and appear to be fairly accurate at least for potentials $U(x)$ that do not vary too rapidly on an atomic scale.

17.4.2   *Two-dimensional conductor*

Related video lecture available at course website, Unit 1: L1.6.

A two-dimensional array of atoms (Fig. 17.7) can be modeled similarly with a nearest neighbor orthogonal TB model, with the model parameters

ε and t chosen to yield a dispersion relation approximating a standard parabolic effective mass relation:

$$E\left(k_x, k_y\right) = E_c + \frac{\hbar^2 \left(k_x^2 + k_y^2\right)}{2m}. \tag{17.17}$$

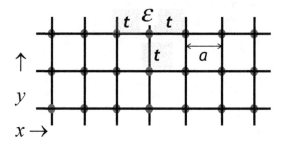

Fig. 17.7   A two-dimensional nearest neighbor orthogonal TB model.

In this case we can assume a solution of the form

$$\psi_n = \psi_0\, e^{i\mathbf{k} \cdot \mathbf{r}_n}$$

where $\mathbf{k} = k_x\, \hat{\mathbf{x}} + k_y\, \hat{\mathbf{y}}$ and $\mathbf{r}_n$ denotes the location of atom n. Substituting into Eq. (17.10) we obtain the dispersion relation

$$E(\mathbf{k}) = \sum_m H_{nm}\, e^{i\mathbf{k} \cdot (\mathbf{r}_m - \mathbf{r}_n)} \tag{17.18a}$$

which for our nearest neighbor model yields

$$E(\mathbf{k}) = \varepsilon + t\, e^{+ik_x a} + t\, e^{-ik_x a} + t\, e^{+ik_y a} + t\, e^{-ik_y a}$$
$$= \varepsilon + 2t\, \cos\left(k_x a\right) + 2t\, \cos\left(k_y a\right). \tag{17.18b}$$

Following the same arguments as in the 1D case, we can make this match the parabolic relation in Eq. (17.17) by choosing

$$t = -\frac{\hbar^2}{2ma^2} \tag{17.19a}$$

and

$$\varepsilon = E_c - 4t. \tag{17.19b}$$

17.4.3 TB parameters in B-field

It is shown in Appendix D that if we replace $\mathbf{p}$ with $\mathbf{p} + q\mathbf{A}$ in Eq. (17.6a)

$$E_{\text{classical}}\left(\mathbf{r}, \mathbf{p}\right) = \frac{\left(\mathbf{p} + q\mathbf{A}\right) \cdot \left(\mathbf{p} + q\mathbf{A}\right)}{2m} + U\left(\mathbf{r}\right)$$

yields the correct classical laws of motion of a particle of charge $-q$ in a vector potential $\mathbf{A}$. The corresponding wave equation is obtained using the replacement in Eq. (17.1): $\mathbf{p} \rightarrow -i\hbar\nabla$.

To find the appropriate TB parameters for the Hamiltonian in a **B**-field we consider the homogeneous material with constant E_c and a constant vector potential. Consider first the 1D problem with

$$E\left(p_x\right) = E_c + \frac{\left(p_x + qA_x\right)\left(p_x + qA_x\right)}{2m}$$

so that the corresponding wave equation has a dispersion relation

$$E\left(k_x\right) = E_c + \frac{\left(\hbar k_x + qA_x\right)\left(\hbar k_x + qA_x\right)}{2m}$$

which can be approximated by a cosine function

$$E\left(k_x\right) = \varepsilon + 2t \cos\left(k_x a + \frac{qA_x a}{\hbar}\right)$$

with ε and t chosen according to Eq. (17.16). This means that we can model it with the 1D lattice shown here, which differs from our original model in Fig. 17.5 by the extra phase $qA_x a/\hbar$.

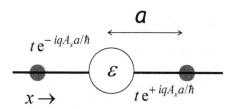

Similar arguments lead to a similar phase in the y-direction as well. This is included in the 2D tight-binding model by modifying the nearest neighbor coupling elements to include an appropriate phase in the nearest neighbor coupling elements as shown in Fig. 17.8 with

$$\varphi_x = \frac{qA_x a}{\hbar}, \qquad \varphi_y = \frac{qA_y a}{\hbar}.$$

To include a **B**-field we have to let the vector potential vary spatially from one lattice point to the next such that

$$\mathbf{B} = \nabla \times \mathbf{A}.$$

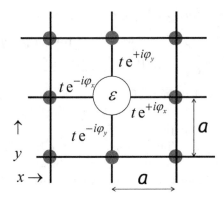

Fig. 17.8   The effect of a magnetic field in the z-direction is included in a tight-binding model by introducing a phase in the nearest neighbor coupling elements as discussed in the text.

For example, a **B**-field in the z-direction described in general by a vector potential $A_x(y)$ and/or $A_y(x)$ such that

$$B_z = \frac{\partial A_y}{\partial x} - \frac{\partial A_x}{\partial y}.$$

For a given **B**-field the potential **A** is not unique, and it is usually convenient to choose a potential that does not vary along the direction of current flow.

17.4.4   *Lattice with a "Basis"*

Related video lecture available at course website, Unit 1: L1.8.

We have seen how for any given TB model we can evaluate the $E(\mathbf{k})$ relation from Eq. (17.18) and then fit it to a desired function. However, Eq. (17.18) will not work if we have a *"lattice with a basis"*. For example if we apply it to the graphene lattice shown in Fig. 17.9, we will get different answers depending on whether we choose "n" to be the left carbon atom or the right carbon atom. The reason is that in a lattice like this these two carbon atoms are not in identical environments: One sees two bonds to the left and one bond to the right, while the other sees one bond to the left and two bonds to the right. We call this a lattice with a basis in the sense that two carbon atoms comprise a **unit cell**: if we view a pair of carbon atoms (marked A and B) as a single entity then the lattice looks *translationally invariant* with each entity in an identical environment.

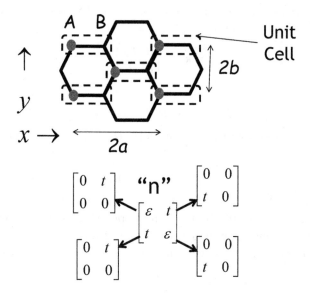

Fig. 17.9 If we view two carbon atoms as a single entity then the lattice in Fig. 17.4 looks translationally invariant with each entity in an identical environment. Viewing the two atoms in each unit cell as a single entity we can write the set of equations in the form shown in Eq. (17.20) with $\mathbf{H}_{nm}$ given by (2×2) matrices as shown.

We can then write the set of equations in Eq. (17.10) in the form

$$E \psi_n = \sum_m \mathbf{H}_{nm} \psi_m \tag{17.20}$$

where ψ_n is a (2×1) column vector whose components represent the two atoms comprising unit cell number n. Similarly $\mathbf{H}_{nm}$ is a (2×2) matrix representing the coupling between the two components of unit cell n and unit cell m (see Fig. 17.9).

Now if we write the solution in the form

$$\psi_n = \psi_0 \, e^{i\mathbf{k} \cdot \mathbf{r}_n} \tag{17.21}$$

we obtain from Eq. (17.20)

$$E \psi_0 = \mathbf{h}(\mathbf{k}) \psi_0 \tag{17.22}$$

where

$$\mathbf{h}(\mathbf{k}) = \sum_m \mathbf{H}_{nm} \, e^{i\mathbf{k} \cdot (\mathbf{r}_m - \mathbf{r}_n)}. \tag{17.23}$$

Note that $\mathbf{h}(\mathbf{k})$ obtained from Eq. (17.23) is a (2×2) matrix, which can be shown after some algebra to be

$$\mathbf{h}(\mathbf{k}) = \begin{bmatrix} \varepsilon & h_0^* \\ h_0 & \varepsilon \end{bmatrix} \tag{17.24}$$

where

$$h_0(\mathbf{k}) \equiv t\left(1 + 2e^{+ik_x a}\cos(k_y b)\right). \tag{17.25}$$

Equation (17.22) yields two eigenvalues for the energy E for each value of $\mathbf{k}$:

$$E(\mathbf{k}) = \varepsilon \pm |h_0(\mathbf{k})|. \tag{17.26}$$

Equations (17.26) and (17.25) give a widely used dispersion relation for graphene. Once again to obtain a simple polynomial relation we need a Taylor series expansion around the k-value of interest. In this case the k-values of interest are those that make

$$h_0(\mathbf{k}) = 0$$

so that

$$E(\mathbf{k}) = \varepsilon.$$

This is because the equilibrium electrochemical potential is located at ε for a neutral sample for which exactly half of all the energy levels given by Eq. (17.26) are occupied.

It is straightforward to see that this requires

$$h_0(\mathbf{k}) = 0 \quad \rightarrow \quad k_x a = 0, \ k_y b = \pm\frac{2\pi}{3}. \tag{17.27}$$

Alternatively one could numerically make a grayscale plot of the magnitude of $h_0(\mathbf{k})$ as shown below and look for the dark spots where it is a minimum. Each of these spots is called a valley and one can do a Taylor expansion around the minimum to obtain an approximate dispersion relation valid for that valley. Note that two of the dark spots correspond to the points in Eq. (17.27), but there are other spots too and it requires some discussion to be convinced that these additional valleys do not need to be considered separately (see for example, Chapter 5, Datta (2005)).

A Taylor expansion around the points in Eq. (17.27) yields

$$h_0(\mathbf{k}) \approx \pm i\,ta\,(k_x \mp i\beta_y), \tag{17.28a}$$

where

$$\beta_y \equiv k_y \mp \frac{2\pi}{3b}. \tag{17.28b}$$

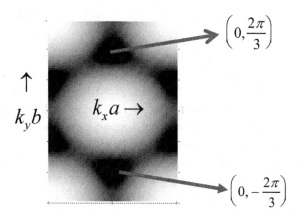

Using this approximate relation we obtain a simple dispersion relation:

$$E = \varepsilon \pm at \sqrt{k_x^2 + k_y^2} \qquad (17.29)$$

which corresponds to the energy-momentum relation

$$E = \nu_0 p$$

that we stated in Chapter 6, if we set $\varepsilon = 0$. The two valleys correspond to the two values of $k_y b$ in Eq. (17.27).

In summary, although the differential form of the Schrödinger equation (Eq. (17.2)) is the well-known one that appears in most textbooks as well as on T-shirts, practical calculations are usually based on a discretized version that represents the Hamiltonian operator, $\mathbf{H}_{op}$ (Eq. (17.8)) as a matrix of size $N \times N$, N being the number of basis functions used to represent the structure.

Given a set of basis functions, the matrix $\mathbf{H}$ can be obtained from first principles, but a widely used approach is to use the principles of bandstructure to represent the matrix in terms of a few parameters which are chosen to match key experiments. Such semi-empirical approaches are often used because of their convenience and can explain a wide range of experiments beyond the key ones used that are used as input, suggesting that they capture a lot of essential physics.

Our approach in this book will be to

(1) take accepted energy-momentum $E(\mathbf{p})$ relations that are believed to describe the dynamics of conduction electrons with energies around the electrochemical potential μ_0,

(2) extract appropriate parameters to use in tight-binding model by discretizing it.

Knowing the $\mathbf{H}$, we can obtain the $\mathbf{\Sigma}_{1,2}$ describing the connection to the physical contacts and possible approaches will be described when discussing specific examples in Chapters 19 through 23. A key difference between the $\mathbf{H}$ and $\mathbf{\Sigma}$ matrices is that the former is Hermitian with real eigenvalues, while the latter is non-Hermitian with complex eigenvalues, whose significance we will discuss in the next chapter.

As we mentioned at the outset, there are many approaches to writing $\mathbf{H}$ of which we have only described the simplest versions. But regardless of how we chose to write these matrices, we can use the NEGF-based approach to be described in the next chapter.

Chapter 18

NEGF Method

In the last chapter I tried to provide a super-brief but hopefully self-contained introduction to the Hamiltonian matrix **H** whose eigenvalues tell us the allowed energy levels in the channel. However, **H** describes an isolated channel and we cannot talk about the steady-state resistance of an isolated channel without bringing in the contacts and the battery connected across it. In this chapter, I will describe the NEGF-based transport model that can be used to model current flow, given **H** and the Σ's (Fig. 18.1).

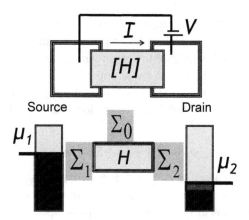

Fig. 18.1 The NEGF-based quantum transport model described here allows us to model current flow given the Hamiltonian matrix **H** describing the channel, the self-energy matrices Σ describing the connection of the channel to the contacts, and Σ_0 describing interactions within the channel.

As I mentioned in Chapters 1 and 17, the NEGF method originated from the seminal works of Martin and Schwinger (1959), Kadanoff and Baym (1962), Keldysh (1965) and others who used the methods of many-body perturbation theory (MBPT) to describe the distributed entropy-generating processes along the channel which were believed to constitute the essence of resistance. Since MBPT is an advanced topic requiring many semesters to master, the NEGF method is generally regarded as an esoteric tool for specialists.

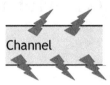

We will start with elastic resistors for which energy exchange is confined to the contacts, and the problem of resistance can be treated within a one-electron picture by connecting contacts to the Schrödinger equation. Indeed our approach will be to start from the usual time-independent Schrödinger equation $E\psi = \mathbf{H}\psi$ and add two terms to it representing the outflow and inflow from the contact

$$E\,\boldsymbol{\psi} \;=\; \mathbf{H}\,\boldsymbol{\psi} \;+\; \underbrace{\boldsymbol{\Sigma}\,\boldsymbol{\psi}}_{\text{OUTFLOW}} \;+\; \underbrace{\mathbf{s}}_{\text{INFLOW}}\;.$$

These two terms arise from imposing open boundary conditions on the Schrödinger equation with an incident wave from the contact as shown in Chapters 8 and 9 of Datta (2005). Some readers may notice the similarity of the additional terms here with those appearing in the *Langevin equation* used to describe *Brownian* motion, but we will not go into it.

Using this modified Schrödinger equation, the wavefunction can be written as

$$\boldsymbol{\psi} \;=\; [E\mathbf{I} - \mathbf{H} - \boldsymbol{\Sigma}]^{-1}\,\mathbf{s}.$$

We will then argue that since the inflow from multiple sources 's' are incoherent, one cannot superpose the resulting $\boldsymbol{\psi}$'s and it is more convenient to work in terms of quantities like (superscript '†' denotes conjugate transpose)

$$\mathbf{G}^{n} \;\sim\; \boldsymbol{\psi}\,\boldsymbol{\psi}^{\dagger}$$

$$\boldsymbol{\Sigma}^{in} \;\sim\; \mathbf{s}\,\mathbf{s}^{\dagger}$$

which can be superposed. Defining

$$\mathbf{G}^{R} \;=\; [E\mathbf{I} - \mathbf{H} - \boldsymbol{\Sigma}]^{-1} \tag{18.1}$$

$$\text{and} \quad \mathbf{G}^A \; = \; \left[\mathbf{G}^R\right]^\dagger$$

we can write

$$\boldsymbol{\psi} \; = \; \mathbf{G}^R \mathbf{s}$$

so that

$$\underbrace{\boldsymbol{\psi}\boldsymbol{\psi}^\dagger}_{\mathbf{G}^n} \; = \; \mathbf{G}^R \underbrace{\mathbf{s}\,\mathbf{s}^\dagger}_{\boldsymbol{\Sigma}^{in}} \mathbf{G}^A$$

giving us the second NEGF equation

$$\mathbf{G}^n \; = \; \mathbf{G}^R \, \boldsymbol{\Sigma}^{in} \, \mathbf{G}^A. \tag{18.2}$$

In this book, we have changed the notation for several key NEGF quantities, writing $\boldsymbol{\Sigma}$ for the "retarded self-energy function" $\boldsymbol{\Sigma}^R$, and more importantly the following symbols (Table 18.1, the reader could also look at Chapter 8, Datta 1995).

Table 18.1 NEGF symbols used in this book and their conventional counterpart used in literature.

Conventional NEGF symbol	Meaning of the symbol	Symbol used in this book
$-i\mathbf{G}^<$	Matrix electron density	$\mathbf{G}^n$
$+i\mathbf{G}^>$	Matrix hole density	$\mathbf{G}^p$
$-i\boldsymbol{\Sigma}^<$	In-scattering function	$\boldsymbol{\Sigma}^{in}$
$+i\boldsymbol{\Sigma}^>$	Out-scattering function	$\boldsymbol{\Sigma}^{out}$

Note that the symbols in the rightmost column are all Hermitian.

Equations (18.1) and (18.2) are essentially the same as Eqs. (75)–(77) in Keldysh (1965), which is one of the seminal founding papers on the NEGF method that obtained these equations using MBPT. Although for simplicity we have only discussed the time-independent version here, a similar derivation could be used for the time-dependent version too (See Appendix, Datta (2005)).

How could we obtain these results using elementary arguments, without invoking MBPT? Because we are dealing with an elastic resistor where all entropy-generating processes are confined to the contacts and can be handled in a relatively elementary manner. But should we call this NEGF?

It seems to us that NEGF has two aspects, namely

A. Eqs. (18.1) and (18.2) and
B. calculating $\boldsymbol{\Sigma}$ and $\boldsymbol{\Sigma}^{in}$ that appear in Eqs. (18.1) and (18.2).

For historical reasons, these two aspects, A and B, are often intertwined in the literature, but they need not be. Indeed these two aspects are completely distinct in the Boltzmann formalism (Chapter 9). The Boltzmann transport equation (BTE)

$$\frac{\partial f}{\partial t} + \boldsymbol{\nu} \cdot \nabla f + \mathbf{F} \cdot \nabla_p f = S_{op} f$$

is used to describe semiclassical transport in many different contexts, but the evaluation of the scattering operator S_{op} has evolved considerably since the days of Boltzmann and varies widely depending on the problem at hand.

Similarly it seems to me that the essence of NEGF is contained in Eqs. (18.1) and (18.2) while the actual evaluation of the Σ's may well evolve as we look at more and more different types of problems. The original MBPT–based approach may or may not be the best, and may need to be modified even for problems involving electron-electron interactions.

Above all we believe that by decoupling Eqs. (18.1) and (18.2) from the MBPT method originally used to derive them, we can make the NEGF method more transparent and accessible so that it can become a part of the standard training of physics and engineering students who need to apply it effectively to a wide variety of basic and applied problems that require connecting contacts to the Schrödinger equation.

I should also note briefly the relation between the NEGF method applied to elastic resistors with the scattering theory of transport or the transmission formalism widely used in mesoscopic physics. Firstly, the scattering theory works directly with the Schrödinger equation with open boundary conditions that effectively add the inflow and outflow terms we mentioned:

$$E \psi = \mathbf{H} \psi + \underbrace{\Sigma \psi}_{\text{OUTFLOW}} + \underbrace{\mathbf{s}}_{\text{INFLOW}} \; .$$

However, as we noted earlier it is then important to add individual sources incoherently, something that the NEGF equation (Eq. (18.2)) takes care of automatically.

The second key difference is the handling of dephasing processes in the channel, something that has no classical equivalent. In quantum transport randomization of the phase of the wavefunction even without any momentum relaxation can have a major impact on the measured conductance. The scattering theory of transport usually neglects such dephasing processes and is restricted to *phase-coherent elastic resistors*.

Incoherence is commonly introduced in this approach using an insightful observation due to Büttiker that dephasing processes essentially remove electrons from the channel and re-inject them just like the voltage probes discussed in Section 10.3 and so one can include them phenomenologically by introducing conceptual contacts in the channel.

This method is widely used in mesoscopic physics, but it seems to introduce both phase and momentum relaxation and I am not aware of a convenient way to introduce pure phase relaxation if we wanted to. In the NEGF method it is straightforward to choose $[\boldsymbol{\Sigma}_0]$ so as to include phase relaxation with or without momentum relaxation as we will see in the next chapter. In addition, the NEGF method provides a rigorous framework for handling all kinds of interactions in the channel, both elastic and inelastic, using MBPT. Indeed that is what the original work from the 1960s was about.

Let me finish up this long introduction by briefly mentioning the two other key equations in NEGF besides Eqs. (18.1) and (18.2). As we will see, the quantity $\mathbf{G}^n$ appearing in Eq. (18.2) represents a matrix version of the electron density (times 2π) from which other quantities of interest can be calculated. Another quantity of interest is the matrix version of the density of states (again times 2π) called the *spectral function* $\mathbf{A}$ given by

$$\mathbf{A} = \mathbf{G}^R \boldsymbol{\Gamma} \mathbf{G}^A = \mathbf{G}^A \boldsymbol{\Gamma} \mathbf{G}^R$$
$$= i\left[\mathbf{G}^R - \mathbf{G}^A\right] = \mathbf{A}^\dagger \tag{18.3a}$$

where $\mathbf{G}^R$ and $\mathbf{G}^A$ are defined in Eq. (18.1) and the $\boldsymbol{\Gamma}$'s represent the *anti-Hermitian* parts of the corresponding $\boldsymbol{\Sigma}$'s

$$\boldsymbol{\Gamma} = i\left[\boldsymbol{\Sigma} - \boldsymbol{\Sigma}^\dagger\right] \tag{18.3b}$$

which describe how easily the electrons in the channel communicate with the contacts.

There is a component of $\boldsymbol{\Sigma}$, $\boldsymbol{\Gamma}$ and $\boldsymbol{\Sigma}^{in}$ for each contact (physical or otherwise) and the quantities appearing in Eqs. (18.1)–(18.3) are the total obtained summing all components. The current at a specific contact m, however, involves only those components associated with contact m:

$$\tilde{I}_m = \frac{q}{h}\,\mathrm{Trace}\left[\boldsymbol{\Sigma}_m^{in}\mathbf{A} - \boldsymbol{\Gamma}_m\,\mathbf{G}^n\right]. \tag{18.4}$$

Note that $\tilde{I}_m(E)$ represents the current per unit energy and has to be integrated over all energy to obtain the total current. In the following four chapters we will look at a few examples designed to illustrate how Eqs. (18.1)–(18.4) are applied to obtain concrete results.

But for the rest of this chapter let me try to justify these equations. We start with a one-level version for which all matrices are just numbers

(Section 18.1), then look at the full multi-level version (Section 18.2), obtain an expression for the conductance function $G(E)$ for coherent transport (Section 18.3) and finally look at the different choices for the dephasing self-energy $\mathbf{\Sigma_0}$ (Section 18.4).

18.1   One-level Resistor

To get a feeling for the NEGF method, it is instructive to look at a particularly simple conductor having just one level and described by a (1×1) $\mathbf{H}$ matrix that is essentially a number: $\mathbf{H} = \epsilon$.

Starting directly from the Schrödinger equation we will see how we can introduce contacts into this problem. This will help set the stage for Section 18.3 when we consider arbitrary channels described by $(N \times N)$ matrices instead of the simple one-level channel described by (1×1) "matrices."

18.1.1   *Semiclassical treatment*

Related video lecture available at course website, Unit 2: L2.2.

It is useful to first go through a semiclassical treatment as an intuitive guide to the quantum treatment. Physically we have a level connected to two contacts, with two different occupancy factors

$$f_1(\varepsilon) \quad \text{and} \quad f_2(\varepsilon).$$

Let us assume the occupation factor to be one for the source and zero for the drain, so that it is only the source that is continually trying to fill up the level while the drain is trying to empty it. We will calculate the resulting current and then multiply it by

$$f_1(\varepsilon) \; - \; f_2(\varepsilon)$$

to account for the fact that there is injection from both sides and the net current is the difference.

With $f_1 = 1$ in the source and $f_2 = 0$ in the drain, the average number N of electrons ($N < 1$) should obey an equation of the form

$$\frac{d}{dt}N \; = \; -(\nu_1 + \nu_2)N + \; S_1 \; + \; S_2 \tag{18.5}$$

where ν_1 and ν_2 represent the rates (per second) at which an electron escapes into the source and drain respectively, while S_1 and S_2 are the

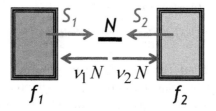

Fig. 18.2 Filling and emptying a level: Semiclassical picture.

rates at which electrons try to enter from the source and drain respectively. The steady state occupation is obtained by setting

$$\frac{d}{dt}N = 0 \quad \rightarrow \quad N = \frac{S_1 + S_2}{\nu_1 + \nu_2}.$$ (18.6)

We can fix S_1, by noting that if the drain were to be disconnected, N should equal the Fermi function $f_1(\varepsilon)$ in contact 1, which we will assume one for this discussion. This means

$$\frac{S_1}{\nu_1} = f_1(\varepsilon) \quad \text{and} \quad \frac{S_2}{\nu_2} = f_2(\varepsilon).$$ (18.7)

The current can be evaluated by writing Eq. (18.5) in the form

$$\frac{dN}{dt} = (S_1 - \nu_1 N) + (S_2 - \nu_2 N)$$ (18.8)

and noting that the first term on the right is the current from the source while the second is the current into the drain. Under steady state conditions, they are equal and either could be used to evaluate the current that flows in the circuit:

$$I = q(S_1 - \nu_1 N) = q(\nu_2 N - S_2).$$ (18.9)

From Eqs. (18.6), (18.7) and (18.9), we have

$$N = \frac{\nu_1 f_1(\varepsilon) + \nu_2 f_2(\varepsilon)}{\nu_1 + \nu_2}$$ (18.10a)

and

$$I = q\frac{\nu_1 \nu_2}{\nu_1 + \nu_2}(f_1(\varepsilon) - f_2(\varepsilon)).$$ (18.10b)

18.1.2   Quantum treatment

Related video lecture available at course website, Unit 2: L2.3.

Let us now work out the same problem using a quantum formalism based on the Schrödinger equation. In the last chapter we introduced the matrix version of the time-independent Schrödinger equation

$$E\psi \;=\; \mathbf{H}\psi$$

which can be obtained from the more general time-dependent equation

$$i\hbar\frac{\partial}{\partial t}\tilde{\boldsymbol{\psi}}(t) \;=\; \mathbf{H}\,\tilde{\boldsymbol{\psi}}(t) \tag{18.11a}$$

by assuming

$$\tilde{\boldsymbol{\psi}}(t) \;=\; \boldsymbol{\psi}\,e^{-iEt/\hbar}. \tag{18.11b}$$

For problems involving steady-state current flow, the time-independent version is usually adequate, but sometimes it is useful to go back to the time-dependent version because it helps us interpret certain quantities like the self-energy functions as we will see shortly.

In the quantum formalism the squared magnitude of the electronic wave-function $\tilde{\psi}(t)$ tells us the probability of finding an electron occupying the level and hence can be identified with the average number of electrons $N(<1)$. For a single isolated level with $\mathbf{H} = \varepsilon$, the time evolution of the wavefunction is described by

$$i\hbar\frac{d}{dt}\tilde{\psi} \;=\; \varepsilon\tilde{\psi}$$

which with a little algebra leads to

$$\frac{d}{dt}\left(\tilde{\psi}\tilde{\psi}^{*}\right) \;=\; 0$$

showing that for an isolated level, the number of electrons $\tilde{\psi}\tilde{\psi}^{*}$ does not change with time.

Our interest, however, is not in isolated systems, but in channels connected to two contacts. Unfortunately the standard quantum mechanics literature does not provide much guidance in the matter, but we can do something relatively simple using the rate equation in Eq. (18.5) as a guide.

We introduce **contacts into the Schrödinger equation** by modifying it to read

$$i\hbar\frac{d}{dt}\tilde{\psi} \;=\; \left(\varepsilon - i\frac{\gamma_1 + \gamma_2}{2}\right)\tilde{\psi} \tag{18.12a}$$

so that the resulting equation for

$$\frac{d}{dt}\tilde{\psi}\tilde{\psi}^* = -\left(\frac{\gamma_1 + \gamma_2}{\hbar}\right)\tilde{\psi}\tilde{\psi}^* \tag{18.12b}$$

looks just like Eq. (18.5) except for the source term S_1 which we will discuss shortly.

We can make Eq. (18.12b) match Eq. (18.5) if we choose

$$\gamma_1 = \hbar\nu_1 \tag{18.13a}$$

$$\gamma_2 = \hbar\nu_2. \tag{18.13b}$$

We can now go back to the ***time-independent version*** of Eq. (18.12a):

$$E\psi = \left(\varepsilon - i\frac{\gamma_1 + \gamma_2}{2}\right)\psi \tag{18.14}$$

obtained by assuming a single energy solution:

$$\tilde{\psi}(t) = \psi(E)\, e^{-iEt/\hbar}.$$

Equation (18.14) has an obvious solution $\psi = 0$, telling us that at steady-state there are no electrons occupying the level which makes sense since we have not included the source term S_1. All electrons can do is to escape into the contacts, and so in the long run the level just empties to zero.

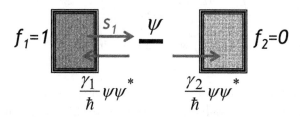

Fig. 18.3   Filling and emptying a level: Quantum picture.

Introducing a source term into Eq. (18.14) and defining $\gamma \equiv \gamma_1 + \gamma_2$, we have

$$E\psi = \left(\varepsilon - i\frac{\gamma}{2}\right)\psi + s_1. \tag{18.15}$$

Unlike the semiclassical case (Eq. (18.5)) we are introducing only one source rather than two. The reason is subtle and we will address it later at the end of this section. From Eq. (18.15), we can relate the wavefunction to the source

$$\psi = \frac{s_1}{E - \varepsilon + i(\gamma/2)}. \qquad (18.16)$$

Note that the wavefunction is a maximum when the electron energy E equals the energy ε of the level, as we might expect. But the important point about the quantum treatment is that the wavefunction is not significantly diminished as long as E differs from ε by an amount less than γ. This is an example of "*broadening*" or energy uncertainty that a semiclassical picture misses.

To obtain the **strength of the source** we require that the total number of electrons on integrating over all energies should equal our rate equation result from Eq. (18.5). That is,

$$\int_{-\infty}^{+\infty} dE \ \psi\psi^* = \frac{\nu_1}{\nu_1 + \nu_2} = \frac{\gamma_1}{\gamma_1 + \gamma_2} \qquad (18.17)$$

where we have made use of Eq. (18.13). We now use Eqs. (18.16) and (18.17) to evaluate the right hand side in terms of the source

$$\int_{-\infty}^{+\infty} dE \ \psi\psi^* = \int_{-\infty}^{+\infty} dE \ \frac{s_1 s_1^*}{(E - \varepsilon)^2 + \left(\dfrac{\gamma}{2}\right)^2} = \frac{2\pi s_1 s_1^*}{\gamma} \qquad (18.18)$$

where we have made use of a standard integral

$$\int_{-\infty}^{+\infty} dE \ \frac{\gamma}{(E - \varepsilon)^2 + \left(\dfrac{\gamma}{2}\right)^2} = 2\pi. \qquad (18.19)$$

From Eqs. (18.17) and (18.18) we obtain, noting that

$$2\pi s_1 s_1^* = \gamma_1. \qquad (18.20)$$

The strength of the source is thus proportional to the escape rate which seems reasonable: if the contact is well coupled to the channel and electrons can escape easily, they should also be able to come in easily.

Just as in the semiclassical case (Eq. (18.9)) we obtain the **current** by looking at the rate of change of N from Eq. (18.12b)

$$\frac{d}{dt}\tilde{\psi}\tilde{\psi}^* = (\text{Inflow from 1}) - \frac{\gamma_1}{\hbar}\tilde{\psi}\tilde{\psi}^* - \frac{\gamma_2}{\hbar}\tilde{\psi}\tilde{\psi}^*$$

where we have added a term "Inflow from 1" as a reminder that Eq. (18.12a) does not include a source term. Both left and right hand sides of this equation are zero for the steady-state solutions we are considering. But just like the semiclassical case, we can identify the current as either the first two terms or the last term on the right:

$$\frac{I}{q} = (\text{Inflow from 1}) - \frac{\gamma_1}{\hbar}\tilde{\psi}\tilde{\psi}^* = \frac{\gamma_2}{\hbar}\tilde{\psi}\tilde{\psi}^*.$$

Using the second form and integrating over energy we can write

$$I = q \int_{-\infty}^{+\infty} dE \, \frac{\gamma_2}{\hbar} \, \psi\psi^* \tag{18.21}$$

so that making use of Eqs. (18.16) and (18.20), we have

$$I = \frac{q}{\hbar} \frac{\gamma_1\gamma_2}{2\pi} \int_{-\infty}^{+\infty} dE \, \frac{1}{(E-\varepsilon)^2 + (\gamma/2)^2} \tag{18.22}$$

which can be compared to the semiclassical result from Eq. (18.10b) with $f_1 = 1$ and $f_2 = 0$ (note: $\gamma = \gamma_1 + \gamma_2$)

$$I = \frac{q}{h} \frac{\gamma_1\gamma_2}{\gamma_1 + \gamma_2}.$$

18.1.3 *Quantum broadening*

Note that Eq. (18.22) involves an integration over energy, as if the quantum treatment has turned the single sharp level into a continuous distribution of energies described by a density of states $D(E)$:

$$D = \frac{\gamma/2\pi}{(E-\epsilon)^2 + (\gamma/2)^2}. \tag{18.23}$$

Quantum mechanically the process of coupling inevitably spreads a single discrete level into a state that is distributed in energy, but integrated over all energy still equals one (see Eq. (18.15)). One could call it a consequence of the **uncertainty relation**

$$\gamma t \geq h$$

relating the length of time t the electron spends in a level to the uncertainty "γ" in its energy. The stronger the coupling, shorter the time and larger the broadening.

Is there any experimental evidence for this energy broadening (Eq. (18.23)) predicted by quantum theory? A Hydrogen molecule has an energy level diagram like the one-level resistor we are discussing and experimentalists have measured the conductance of a Hydrogen molecule with good contacts and it supports the quantum result (Smit *et al.*, 2002). Let me elaborate a little.

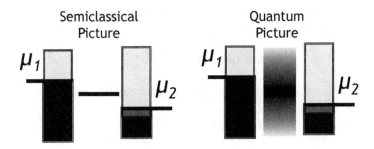

Comparing Eq. (18.22) with Eq. (3.3) for elastic resistors we can write the conductance function for a one-level device including quantum broadening as

$$G(E) = \frac{q^2}{h} \frac{\gamma_1 \gamma_2}{(E - \varepsilon)^2 + \left(\frac{\gamma}{2}\right)^2}.$$

If we assume **(1)** equal coupling to both contacts:

$$\gamma_1 = \gamma_2 = \frac{\gamma}{2}$$

and **(2)** a temperature low enough that the measured conductance equals $G(E = \mu_0)$, μ_0 being the equilibrium electrochemical potential, we have

$$G \approx G(E = \mu_0) = \frac{q^2}{h} \cdot \frac{(\gamma/2)^2}{(\mu_0 - \varepsilon)^2 + (\gamma/2)^2}.$$

So the quantum theory of the one-level resistor says that the measured conductance should show a maximum value equal to the quantum of conductance q^2/h when μ_0 is located sufficiently close to ε. The experimentally measured conductance is equal to $2q^2/h$, the extra factor of 2 being due to *spin degeneracy*, since levels come in pairs and what we have is really a two-level rather than a one-level resistor.

18.1.4 *Do multiple sources interfere?*

In our quantum treatment we considered a problem with electrons injected only from the source ($f_1 = 1$) with the drain empty ($f_2 = 0$) (Eq. (18.15)), unlike the semiclassical case where we started with both sources S_1 and S_2 (Eq. (18.5)).

This is not just a matter of convenience. If instead of Eq. (18.15) we start from

$$E \psi = \left(\varepsilon - i\frac{\gamma}{2}\right) \psi + s_1 + s_2$$

we obtain

$$\psi = \frac{s_1 + s_2}{E - \varepsilon + i\,(\gamma/2)}$$

so that

$$\psi\psi^* = \frac{1}{(E - \varepsilon)^2 + (\gamma/2)^2}\,(s_1 s_1^* + s_2 s_2^* + \underbrace{s_1 s_2^* + s_2 s_1^*}_{\text{Interference} \\ \text{Terms}})$$

which has two extra interference terms that are never observed experimentally because the electrons injected from separate contacts have uncorrelated phases that change randomly in time and average to zero.

The first two terms on the other hand add up since they are positive numbers. It is like adding up the light from two light bulbs: we add their powers not their electric fields. Laser sources on the other hand can be coherent so that we actually add electric fields and the interference terms can be seen experimentally. Electron sources from superconducting contacts too can be coherent leading to *Josephson current* that depend on interference. But that is a different matter.

Our point here is simply that normal contacts like the ones we are discussing are incoherent and it is necessary to take that into account in our models. The moral of the story is that we cannot just insert multiple sources into the Schrödinger equation. We should insert one source at a time, calculate bilinear quantities (things that depend on the product of wavefunctions) like electron density and current and add up the contributions from different sources. Next we will describe the non-equilibrium Green's function (NEGF) method that allows us to implement this procedure in a systematic way and also to include incoherent processes.

18.2 Quantum Transport Through Multiple Levels

We have seen how we can treat quantum transport through a one-level resistor with a time-independent Schrödinger equation modified to include the connection to contacts and a source term:

$$E\,\psi = \left(\varepsilon - i\frac{\gamma}{2}\right)\psi + s.$$

How do we extend this method to a more general channel described by an $N \times N$ Hamiltonian matrix $\mathbf{H}$ whose eigenvalues give the N energy levels?

For an N-level channel, the wavefunction ψ and source term s_1 are $N \times 1$ column vectors and the modified Schrödinger equation looks like

$$E\psi = [\mathbf{H} + \mathbf{\Sigma}_1 + \mathbf{\Sigma}_2]\,\psi + s_1 \qquad (18.24)$$

where $\mathbf{\Sigma}_1$ and $\mathbf{\Sigma}_2$ are $N \times N$ non-Hermitian matrices whose anti-Hermitian components

$$\mathbf{\Gamma}_1 = i\left[\mathbf{\Sigma}_1 - \mathbf{\Sigma}_1^\dagger\right]$$

$$\mathbf{\Gamma}_2 = i\left[\mathbf{\Sigma}_2 - \mathbf{\Sigma}_2^\dagger\right]$$

play the roles of $\gamma_{1,2}$ in our one-level problem.

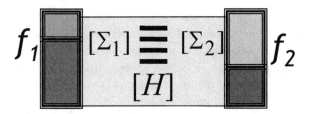

Fig. 18.4   Transport model for multi-level conductor.

In Chapter 17 we discussed how for different structures we can write down the channel Hamiltonian $\mathbf{H}$ and in the next few chapters I will present examples to show how the $\mathbf{\Sigma}$ are obtained.

For the moment, let us focus on how the basic NEGF equations summarized earlier (Eqs. (18.1)–(18.4)) follow from our contact-ed Schrödinger equation, Eq. (18.24).

18.2.1   *Obtaining Eqs. (18.1)*

Related video lecture available at course website, Unit 2: L2.4.

From Eq. (18.24) it is straightforward to write

$$\psi = \mathbf{G}^R s_1$$

where $\mathbf{G}^R$ is given by Eq. (18.1) with

$$\mathbf{\Sigma} = \mathbf{\Sigma}_1 + \mathbf{\Sigma}_2. \qquad (18.25)$$

18.2.2 *Obtaining Eqs. (18.2)*

The matrix electron density, $\mathbf{G}^n$, defined as

$$\mathbf{G}^n \;\rightarrow\; 2\pi\,\boldsymbol{\psi}\,\boldsymbol{\psi}^\dagger \;=\; 2\pi\,\mathbf{G}^R\,\mathbf{s}_1\mathbf{s}_1^\dagger\,\mathbf{G}^A$$

where the superscript "$\dagger$" stands for conjugate transpose, and $\mathbf{G}^A$ stands for the conjugate transpose of $\mathbf{G}^R$.

For the one-level problem $2\pi s_1 s_1^* \;=\; \gamma_1$ (see Eq. (18.20)): the corresponding matrix relation is

$$2\pi\,\mathbf{s}_1\mathbf{s}_1^\dagger \;=\; \boldsymbol{\Gamma}_1$$

so that

$$\mathbf{G}^n \;=\; \mathbf{G}^R\,\boldsymbol{\Gamma}_1\,\mathbf{G}^A.$$

This is for a single source term. For multiple sources, the electron density matrices, unlike the wavefunctions, can all be added up with the appropriate Fermi function weighting to give Eq. (18.2),

$$\mathbf{G}^n \;=\; \mathbf{G}^R\,\boldsymbol{\Sigma}^{in}\,\mathbf{G}^A \quad \text{(same as Eq. (18.2))}$$

with $\boldsymbol{\Sigma}^{in}$ representing an incoherent sum of all the independent sources:

$$\boldsymbol{\Sigma}^{in} \;=\; \boldsymbol{\Gamma}_1\,f_1(E) \;+\; \boldsymbol{\Gamma}_2\,f_2(E). \tag{18.26}$$

18.2.3 *Obtaining Eq. (18.3)*

Equation (18.2) gives us the electron density matrix $\mathbf{G}^n$, in terms of the Fermi functions f_1 and f_2 in the two contacts. But if both f_1 and f_2 are equal to one then all states are occupied, so that the matrix electron density becomes equal to the matrix density of states, called the *spectral function matrix* $\mathbf{A}$. Setting $f_1 = 1$ and $f_2 = 1$ in Eq. (18.26) yields $\boldsymbol{\Sigma}^{in} = \boldsymbol{\Gamma}_1 + \boldsymbol{\Gamma}_2$ and plugging it in Eq. (18.2) we have

$$\mathbf{A} \;=\; \mathbf{G}^R\,\boldsymbol{\Gamma}\,\mathbf{G}^A \tag{18.27}$$

since $\boldsymbol{\Gamma} = \boldsymbol{\Gamma}_1 + \boldsymbol{\Gamma}_2$. This gives us part of Eq. (18.3). The rest of Eq. (18.3) can be obtained from Eq. (18.1) using straightforward algebra as follows:

$$\left[\mathbf{G}^R\right]^{-1} \;=\; E\,\mathbf{I} - \mathbf{H} - \boldsymbol{\Sigma}. \tag{18.28a}$$

Taking conjugate transpose of both sides

$$\left[[\mathbf{G}^R]^{-1}\right]^\dagger \;=\; \left[[\mathbf{G}^R]^\dagger\right]^{-1} \;=\; E\,\mathbf{I} - \mathbf{H} - \boldsymbol{\Sigma}^\dagger. \tag{18.28b}$$

Subtracting Eq. (18.28b) from (18.28a) (note that $\mathbf{G}^A$ stands for $[\mathbf{G}^R]^\dagger$) and making use of Eq. (18.3b)

$$[\mathbf{G}^R]^{-1} - [\mathbf{G}^A]^{-1} = i\,\boldsymbol{\Gamma}. \qquad (18.28c)$$

Multiplying Eq. (18.28c) with $\mathbf{G}^R$ from the left and $\mathbf{G}^A$ from the right we have

$$i\left[\mathbf{G}^R - \mathbf{G}^A\right] = \mathbf{G}^R\,\boldsymbol{\Gamma}\,\mathbf{G}^A$$

thus giving us another piece of Eq. (18.3). The final piece is obtained by multiplying Eq. (18.28c) with $\mathbf{G}^A$ from the left and $\mathbf{G}^R$ from the right.

18.2.4   *Obtaining Eq. (18.4): the current equation*

Related video lecture available at course website, Unit 2: L2.5.

Like the semiclassical treatment and the one-level quantum treatment, the current expression is obtained by considering the time variation of the number of electrons N. Starting from

$$i\hbar\,\frac{d}{dt}\,\psi = [\mathbf{H}+\boldsymbol{\Sigma}]\,\psi + \mathbf{s}$$

and its conjugate transpose (noting that $\mathbf{H}$ is a Hermitian matrix)

$$-i\hbar\,\frac{d}{dt}\,\psi^\dagger = \psi^\dagger\left[\mathbf{H}+\boldsymbol{\Sigma}^\dagger\right] + \mathbf{s}^\dagger$$

we can write

$$i\hbar\,\frac{d}{dt}\psi\psi^\dagger = \left(i\hbar\,\frac{d}{dt}\,\psi\right)\psi^\dagger + \psi\left(i\hbar\,\frac{d}{dt}\,\psi^\dagger\right)$$

$$= \left([\mathbf{H}+\boldsymbol{\Sigma}]\,\psi+\mathbf{s}\right)\psi^\dagger - \psi\left(\psi^\dagger\left[\mathbf{H}+\boldsymbol{\Sigma}^\dagger\right]+\mathbf{s}^\dagger\right)$$

$$= \left[(\mathbf{H}+\boldsymbol{\Sigma})\,\psi\psi^\dagger - \psi\psi^\dagger(\mathbf{H}+\boldsymbol{\Sigma}^\dagger)\right] + \left[\mathbf{s}\mathbf{s}^\dagger\mathbf{G}^A - \mathbf{G}^R\mathbf{s}\mathbf{s}^\dagger\right]$$

where we have made use of the relations

$$\psi = \mathbf{G}^R\mathbf{s} \quad \text{and} \quad \psi^\dagger = \mathbf{s}^\dagger\,\mathbf{G}^A.$$

Since the trace of $\psi\psi^\dagger$ represents the number of electrons, we could define its time derivative as a matrix current operator whose trace gives us the current. Noting further that

$$2\pi\,\psi\psi^\dagger = \mathbf{G}^n \quad \text{and} \quad 2\pi\,\mathbf{s}\mathbf{s}^\dagger = \boldsymbol{\Gamma}$$

we can write

$$I^{op} = \frac{[\mathbf{H}\mathbf{G}^n - \mathbf{G}^n\mathbf{H}] + [\boldsymbol{\Sigma}\mathbf{G}^n - \mathbf{G}^n\boldsymbol{\Sigma}^\dagger] + [\boldsymbol{\Sigma}^{in}\mathbf{G}^A - \mathbf{G}^R\boldsymbol{\Sigma}^{in}]}{i\,2\pi\hbar}.$$

$$(18.29)$$

We will talk more about the current operator in Chapter 23 when we talk about spins, but for the moment we just need its trace which tells us the time rate of change of the number of electrons N in the channel

$$\frac{dN}{dt} = -\frac{i}{h} \, \text{Trace} \left([\mathbf{\Sigma}\mathbf{G}^n - \mathbf{G}^n\mathbf{\Sigma}^\dagger] + [\mathbf{\Sigma}^{in}\mathbf{G}^A - \mathbf{G}^R\mathbf{\Sigma}^{in}] \right)$$

noting that $\text{Trace}\,[\mathbf{AB}] = \text{Trace}\,[\mathbf{BA}]$. Making use of Eq. (18.3b)

$$\frac{dN}{dt} = \frac{1}{h} \, \text{Trace} \left[\mathbf{\Sigma}^{in}\mathbf{A} - \mathbf{\Gamma}\mathbf{G}^n \right].$$

Now comes a tricky argument. Both the left and the right hand sides of Eq. (18.29) are zero, since we are discussing steady state transport with no time variation. The reason we are spending all this time discussing something that is zero is that the terms on the left can be separated into two parts, one associated with contact 1 and one with contact 2. They tell us the currents at contacts 1 and 2 respectively and the fact that they add up to zero is simply a reassuring statement of *Kirchhoff's law* for steady-state currents in circuits.

With this in mind we can write for the current at contact m $(m = 1, 2)$

$$\tilde{I}_m = \frac{q}{h} \, \text{Trace} \left[\mathbf{\Sigma}^{in}_m\mathbf{A} - \mathbf{\Gamma}_m\mathbf{G}^n \right]$$

as stated earlier in Eq. (18.4). This leads us to the picture shown in Fig. 18.5 where we have also shown the semiclassical result for comparison.

18.3 Conductance Functions for Coherent Transport

Finally we note that using Eqs. (18.2)–(18.3) we can write the current from Eq. (18.4) a little differently

$$\tilde{I}(E) = \frac{q}{h} \, \text{Trace} \left[\mathbf{\Gamma}_1\mathbf{G}^R\mathbf{\Gamma}_2\mathbf{G}^A \right] (f_1(E) - f_2(E))$$

which is very useful for it suggests a quantum expression for the conductance function $G(E)$ that we introduced in Chapter 3 for all elastic resistors:

$$G(E) = \frac{q^2}{h} \, \text{Trace} \left[\mathbf{\Gamma}_1\mathbf{G}^R\mathbf{\Gamma}_2\mathbf{G}^A \right]. \tag{18.30}$$

More generally with multiterminal conductors we could introduce a self-energy function for each contact and show that

$$\tilde{I}_m = \frac{q}{h} \sum_n \bar{T}_{mn} (f_m(E) - f_n(E)) \tag{18.31}$$

with

$$\bar{T}_{mn} \equiv \text{Trace} \left[\mathbf{\Gamma}_m\mathbf{G}^R\mathbf{\Gamma}_n\mathbf{G}^A \right]. \tag{18.32}$$

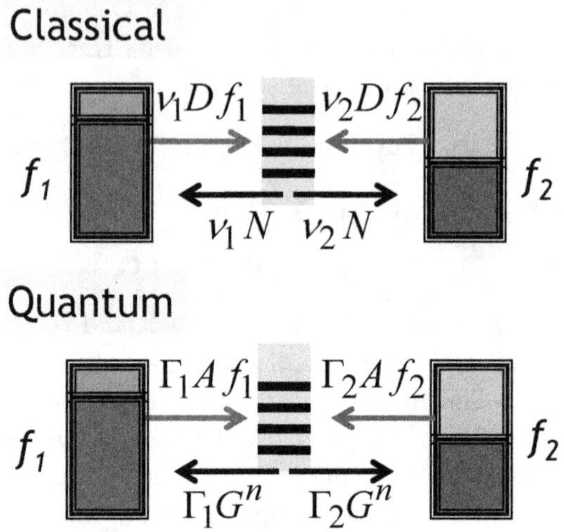

Classical

Quantum

Fig. 18.5   Filling and emptying a channel: Classical and quantum treatment.

For low bias we can use our usual Taylor series expansion from Eq. (2.11) to translate the Fermi functions into electrochemical potentials so that Eq. (18.31) looks just like the Büttiker equation (Eq. (10.3)) with the conductance function given

$$G_{m,n}(E) \equiv \frac{q^2}{h} \text{Trace} \left[\mathbf{\Gamma}_m \mathbf{G}^R \mathbf{\Gamma}_n \mathbf{G}^A \right] \qquad (18.33)$$

which is energy-averaged in the usual way for elastic resistors (see Eq. (3.1)).

$$G_{m,n} = \int_{-\infty}^{+\infty} dE \left(-\frac{\partial f_0}{\partial E} \right) G_{m,n}(E).$$

18.4   Elastic Dephasing

Related video lecture available at course website, Unit 2: L2.9.

So far we have focused on the physical contacts described by $\mathbf{\Sigma}_{1,2}$ and the model as it stands describes coherent quantum transport where electrons travel coherently from source to drain in some static structure described by the Hamiltonian $\mathbf{H}$ without any interactions along the channel described by $\mathbf{\Sigma}_0$ (Fig. 18.1). In order to include $\mathbf{\Sigma}_0$, however, no change is needed as far as Eqs. (18.1) through (18.4) is concerned. It is just that an additional term appears in the definition of Σ and Σ^{in}:

$$\Sigma = \Sigma_1 + \Sigma_2 + \Sigma_0$$
$$\Gamma = \Gamma_1 + \Gamma_2 + \Gamma_0$$
$$\Sigma^{in} = \Gamma_1 f_1(E) + \Gamma_2 f_2(E) + \Sigma_0^{in}. \tag{18.34}$$

What does Σ_0 represent physically? From the point of view of the electron a solid does not look like a static medium described by $\mathbf{H}$, but like a rather turbulent medium with a random potential U_R that fluctuates on a picosecond time scale. Even at fairly low temperatures when phonons have been frozen out, an individual electron continues to see a fluctuating potential due to all the other electrons, whose average is modeled by the SCF potential we discussed in Section 17.2. These fluctuations do not cause any overall loss of momentum from the system of electrons, since any loss from one electron is picked up by another. However, they do cause fluctuations in the phase leading to fluctuations in the current. What typical current measurements tell us is an average flow over nanoseconds if not microseconds or milliseconds. This averaging effect needs to be modeled if we wish to relate to experiments.

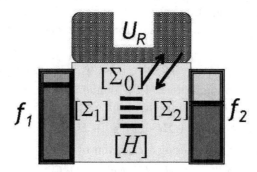

Fig. 18.6   Quantum transport model with simple elastic dephasing.

As we mentioned earlier, the NEGF method was originally developed in the 1960s to deal with the problem of including inelastic processes into a quantum description of large conductors. For the moment, however, we will look at simple elastic dephasing processes leaving more general interactions for Chapter 22.

For such processes the self-energy functions are given by

$$\Sigma_0 = \mathbf{D} \times \mathbf{G}^R \tag{18.35a}$$

$$\mathbf{\Sigma}_0^{in} = \mathbf{D} \times \mathbf{G}^n \qquad (18.35b)$$

where $\times$ denotes element by element multiplication. Making use of the relations in Eqs. (18.3), it is straightforward to show from Eq. (18.35a) that

$$\mathbf{\Gamma}_0 = \mathbf{D} \times \mathbf{A}. \qquad (18.35c)$$

The elements of the matrix $\mathbf{D}$ represent the correlation between the random potential at location "i" and at location "j":

$$D_{ij} = \langle U_{Ri} U_{Rj} \rangle. \qquad (18.36)$$

Two cases are of particular interest. The first is where the random potential is well-correlated throughout the channel having essentially the same value at all points "i" so that the every element of the matrix $\mathbf{D}$ has essentially the same value D_0:

$$\textbf{\textit{Model A:}} \quad D_{ij} = D_0. \qquad (18.37)$$

The other case is where the random potential has zero correlation from one spatial point i to another j, so that

$$\textbf{\textit{Model B:}} \ D_{ij} = \begin{cases} D_0 & \text{if } i = j \\ 0 & \text{if } i \neq j \end{cases}. \qquad (18.38)$$

Real processes are usually somewhere between the two extremes represented by models A and B.

To see where Eqs. (18.35) come from we go back to our contact-ed Schrödinger equation

$$E \, \psi = [\mathbf{H} + \mathbf{\Sigma}_1 + \mathbf{\Sigma}_2] \, \psi + \mathbf{s}_1$$

and noting that a random potential U_R should lead to an additional term that could be viewed as an additional source term

$$E \, \psi = [\mathbf{H} + \mathbf{\Sigma}_1 + \mathbf{\Sigma}_2] \, \psi + U_R \psi + \mathbf{s}_1$$

with a corresponding inscattering term given by

$$\mathbf{\Sigma}_0^{in} = 2\pi \, U_R U_R^* \, \psi \psi^\dagger = D_0 \, \mathbf{G}^n$$

corresponding to Model A (Eq. (18.37)) and a little more careful argument leads to the more general result in Eq. (18.36). That gives us Eq. (18.35b). How about Eq. (18.35a) and (18.35c)?

The simplest way to justify Eq. (18.35c) is to note that together with Eq. (18.35b) (which we just obtained) it ensures that the current at terminal 0 from Eq. (18.4) equals zero:

$$\begin{aligned} I_0 &= \frac{q}{h} \, \text{Trace} \, [\mathbf{\Sigma}_0^{in} \mathbf{A} - \mathbf{\Gamma}_0 \mathbf{G}^n] \\ &= \frac{q}{h} \, \text{Trace} \, [\mathbf{G}^n \mathbf{\Gamma}_0 - \mathbf{\Gamma}_0 \mathbf{G}^n] = 0. \end{aligned}$$

This is a required condition since terminal 0 is not a physical contact where electrons can actually exit or enter from.

Indeed a very popular method due to Büttiker introduces incoherent processes by including a fictitious probe (often called a Büttiker probe) whose electrochemical potential is adjusted to ensure that it draws zero current. In NEGF language this amounts to assuming

$$\Sigma_0^{in} = \mathbf{\Gamma}_0 \, f_P$$

with the number f_P is adjusted for zero current. This would be equivalent to the approach described here if the probe coupling $\mathbf{\Gamma}_0$ were chosen proportional to the spectral function $\mathbf{A}$ as required by Eq. (18.35c).

Note that our prescription in Eq. (18.35) requires a "self-consistent evaluation" since $\mathbf{\Sigma}$ and $\mathbf{\Sigma}^{in}$ depend on $\mathbf{G}^R$ and $\mathbf{G}^n$ which in turn depend on $\mathbf{\Sigma}$ and $\mathbf{\Sigma}^{in}$ respectively (see Eqs. (18.1) and (18.2)).

Also, Model A (Eq. (18.37)) requires us to calculate the full Green's function which can be numerically challenging for large devices described by large matrices. Model B makes the computation numerically much more tractable because one only needs to calculate the diagonal elements of the Green's functions which can be done much faster using powerful algorithms.

In this book, however, we focus on conceptual issues using "toy" problems for which numerical issues are not the "show stoppers." The important conceptual distinction between Models A and B is that the former destroys phase but not momentum, while the latter destroys momentum as well [Golizadeh-Mojarad *et al.*, 2007].

The dephasing process can be viewed as extraction of the electron from a state described by $\mathbf{G}^n$ and reinjecting it in a state described by $\mathbf{D} \times \mathbf{G}^n$. Model A is equivalent to multiplying $\mathbf{G}^n$ by a constant so that the electron is reinjected in exactly the same state that it was extracted in, causing no loss of momentum, while Model B throws away the off-diagonal elements and upon reinjection the electron is as likely to go on way or another. Hopefully this will get clearer in the next chapter when we look at a concrete example.

Another question that the reader might raise is whether instead of including elastic dephasing through a self-energy function $\mathbf{\Sigma}_0$ we could include a potential U_R in the Hamiltonian itself and then average over a number of random realizations of U_R. The answer is that the two methods are not exactly equivalent though in some problems they could yield similar results. This too should be a little clearer in the next chapter when we look at a concrete example.

For completeness, let me note that in the most general case D_{ijkl} is a *fourth order tensor* and the version we are using (Eqs. (18.35)) represents a special case for which D_{ijkl} is non-zero only if $i = k, j = l$ (see Appendix G).

Chapter 19

Can Two Offer Less Resistance than One?

In the next three chapters we will go through a few examples of increasing complexity which are interesting in their own right but have been chosen primarily as "do it yourself" problems that the reader can use to get familiar with the quantum transport model outlined in the last chapter. The MATLAB codes are all included in Appendix H.

In this chapter, we will use 1D quantum transport models to study an interesting question regarding multiple scatterers or obstacles along a conductor. Are we justified in neglecting all interference effects among them and assuming that electrons diffuse like classical particles as we do in the semiclassical picture?

This was the question Anderson raised in his 1958 paper entitled "Absence of Diffusion in Certain Random Lattices" pointing out that diffusion could be slowed significantly and even suppressed completely due to quantum interference between scatterers. "Anderson localization" is a vast topic and we are only using some related issues here to show how the NEGF model provides a convenient conceptual framework for studying interesting physics.

For any problem we need to discuss how we write down the Hamiltonian **H** and the contact self-energy matrices **Σ**. Once we have these, the computational process is standard. The rest is about understanding and enjoying the physics.

19.1 Modeling 1D Conductors

Related video lecture available at course website, Unit 2: L2.6.

For the one-dimensional examples discussed in this chapter, we use the 1D Hamiltonian from Fig. 17.5, shown here in Fig. 19.1. As we discussed earlier for a uniform wire the dispersion relation is given by

$$E(k) = \varepsilon + 2t \cos(ka) \tag{19.1a}$$

which can approximate a parabolic dispersion

$$E = E_c + \frac{\hbar^2 k^2}{2m} \tag{19.1b}$$

by choosing

$$E_c = \varepsilon + 2t \tag{19.2a}$$

$$\text{and} \quad -t \equiv t_0 \equiv \frac{\hbar^2}{2ma^2}. \tag{19.2b}$$

It is straightforward to write down the $\mathbf{H}$ matrix with ε on the diagonal and "t" on the upper and lower diagonals. What needs discussion are the **self-energy matrices**. The basic idea is to replace an infinite conductor described by the Hamiltonian $\mathbf{H}$ with a finite conductor described by $[\mathbf{H} + \boldsymbol{\Sigma}_1 + \boldsymbol{\Sigma}_2]$ assuming **open boundary conditions** at the ends, which means that electron waves escaping from the surface do not give rise to any reflected waves, as a good contact should ensure.

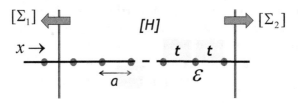

Fig. 19.1   For the one-dimensional examples discussed in this chapter, we use the 1D Hamiltonian from Fig. 17.5.

For a one-dimensional lattice the idea is easy to see. We start from the original equation for the extended system

$$E \psi_n = t\psi_{n-1} + \varepsilon \psi_n + t \psi_{n+1}$$

and then assume that the contact has no incoming wave, just an outgoing wave, so that we can write

$$\psi_{n+1} = \psi_n e^{ika}$$

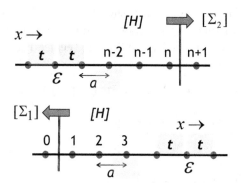

which gives

$$E\,\psi_n \;=\; t\,\psi_{n-1} + \left(\varepsilon + t\,e^{ika}\right)\psi_n.$$

In other words the effect of the contact is simply to add $t\exp\left(+ika\right)$ to H_{nn} which amounts to adding the self-energy

$$\Sigma_1 \;=\; \begin{bmatrix} te^{ika} & 0 & 0 & \cdots & 0 \\ 0 & 0 & 0 & \cdots & 0 \\ 0 & 0 & 0 & \cdots & 0 \\ \vdots & \vdots & \vdots & \ddots & \vdots \\ 0 & 0 & 0 & \cdots & 0 \end{bmatrix}$$

to the Hamiltonian. Note the only non-zero element is the (1,1) element. Similarly at the other contact we obtain

$$\Sigma_2 \;=\; \begin{bmatrix} 0 & \cdots & 0 & 0 & 0 \\ \vdots & \ddots & \vdots & \vdots & \vdots \\ 0 & \cdots & 0 & 0 & 0 \\ 0 & \cdots & 0 & 0 & 0 \\ 0 & \cdots & 0 & 0 & te^{ika} \end{bmatrix}.$$

Note that the only non-zero element is the (n, n) element.

In short, the self-energy function for each contact has a single non-zero element corresponding to the point that is connected to that contact.

19.1.1 *1D ballistic conductor*

Related video lecture available at course website, Unit 2: L2.7.

A good test case for any theory of coherent quantum transport is the conductance function for a length of uniform ballistic conductor: If we are doing things right, the conductance function $G(E)$ should equal the quantum of conductance q^2/h times an integer equal to the number of modes $M(E)$ which is one for 1D conductors (neglecting spin). This means that the transmission (see Eq. (18.32))

$$\bar{T}(E) = \text{Trace} \left[\mathbf{\Gamma}_1 \mathbf{G}^R \mathbf{\Gamma}_2 \mathbf{G}^A \right] \qquad (19.3)$$

should equal one over the energy range

$$0 < E - E_c < 4t_0$$

covered by the dispersion relation

$$E = \varepsilon + 2t \, \cos{(ka)} = E_c + 2t_0(1 - \cos{(ka)}) \qquad (19.4)$$

but zero outside this range (see Fig. 19.2 below with $U = 0$). This is a relatively simple but good example to try to implement numerically when getting started. Obtaining a constant conductance across the entire band is usually a good indicator that the correct self-energy functions are being used and things have been properly set up.

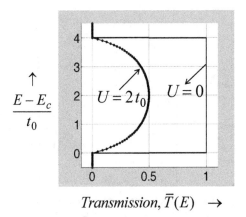

Transmission, $\overline{T}(E) \;\to$

Fig. 19.2   Transmission through a single point scatterer in a 1D wire. For MATLAB script, see Appendix H.1.1.

19.1.2   *1D conductor with one scatterer*

Another good example is that of a conductor with just one scatterer whose effect is included in the Hamiltonian **H** by changing the diagonal element corresponding to that point to $\varepsilon + U$:

$$\mathbf{H} \;=\; \begin{bmatrix} \ddots & \vdots & \vdots & \vdots & \ddots \\ \cdots & \varepsilon & t & 0 & \cdots \\ \cdots & t & \varepsilon + U & t & \cdots \\ \cdots & 0 & t & \varepsilon & \cdots \\ \ddots & \vdots & \vdots & \vdots & \ddots \end{bmatrix}$$

Fig. 19.2 shows the numerical results for $U = 0$ (ballistic conductor) and for $U = 2\,t_0$. Actually there is a simple analytical expression for the transmission through a single point scatterer

$$\bar{T}(E) \;=\; \frac{(2t\,\sin{(ka)})^2}{U^2 + (2t\,\sin{(ka)})^2} \;=\; \frac{(\hbar v/a)^2}{U^2 + (\hbar v/a)^2} \tag{19.5}$$

that we can use to check our numerical results. This expression is obtained by treating the single point where the scatterer is located as the channel, so that all matrices in the NEGF method are (1×1) matrices, that is, just numbers:

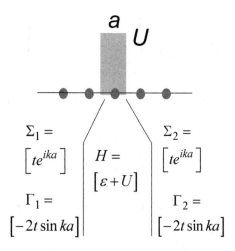

It is easy to see that the Green's function is given by

$$\mathbf{G}^R(E) \;=\; \frac{1}{E - (\varepsilon + U) - 2t\,e^{ika}} \;=\; \frac{1}{-U - i2t\,\sin{(ka)}}$$

making use of Eq. (19.2). Hence

$$\mathbf{\Gamma}_1 \mathbf{G}^R \mathbf{\Gamma}_2 \mathbf{G}^A \;=\; \frac{(2t\,\sin{(ka)})^2}{U^2 + (2t\,\sin{(ka)})^2}$$

giving us the stated result in Eq. (19.5). The second form is obtained by noting from Eq. (19.1a) that

$$\hbar\nu = \frac{dE}{dk} = -2at\sin(ka). \tag{19.6}$$

Once you are comfortable with the results in Fig. 19.2 and are able to reproduce it, you should be ready to include various potentials into the Hamiltonian and reproduce the rest of the examples in this chapter.

19.2 Quantum Resistors in Series

Related video lecture available at course website, Unit 2: L2.8.

In Chapter 10 we argued that the resistance of a conductor with one scatterer with a transmission probability T can be divided into a scatterer resistance and an interface resistance (see Eqs. (10.1) and (10.2))

$$R_1 = \frac{h}{q^2 M} \left(\underbrace{\frac{1-T}{T}}_{\text{scatterer}} + \underbrace{1}_{\text{interface}} \right).$$

What is the resistance if we have two scatterers each with transmission T?

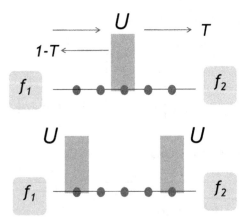

Fig. 19.3   Channel with (a) one scatterer, and (b) two scatterers.

We would expect the scatterer contribution to double:

$$R_2 = \frac{h}{q^2 M} \left(2 \underbrace{\frac{1-T}{T}}_{\text{scatterer}} + \underbrace{1}_{\text{interface}} \right)$$

$$= \frac{h}{q^2 M} \frac{2-T}{T}.$$

We can relate the two resistances by the relation:

$$R_2 = R_1 (2 - T).$$

If T is close to one we have the ballistic limit with $R_2 = R_1$: two sections in series have the same resistance as one of them, since all the resistance comes from the interfaces.

If $T \ll 1$, we have the Ohmic limit with $R_2 = 2R_1$: two sections have twice the resistance as one of them, since all the resistance comes from the channel.

But can R_2 ever be less than R_1? Not as long as electrons behave like classical particles. There is no way an extra roadblock on a classical highway can increase the traffic flow. But on a quantum highway this is quite possible due to wave interference.

We could use our 1D model to study problems of this type. Figure 19.4 shows the transmission functions $\bar{T}(E)$ calculated numerically for a conductor with one scatterer and a conductor with two scatterers.

If the electrochemical potential happens to lie at an energy like the one marked "B", R_2 will be even larger than the Ohmic result R_1. But if the electrochemical potential lies at an energy like the one marked "A", R_2 is less than R_1.

At such energies, the presence of the second scatterer creates a reflection that cancels the reflection from the first one, because they are spaced a quarter wavelength apart. Such quarter wave sections are widely used to create anti-reflection coatings on optical lenses and are well-known in the world of waves, though they are unnatural in the world of particles.

Actually there is a class of devices called resonant tunneling diodes that deliberately engineer two strategically spaced barriers and make use of the resulting sharp peaks in conductance to achieve interesting current-voltage characteristics like the one sketched here where over a range of voltages, the slope dI/dV is negative ("negative differential resistance, NDR"). We could use our elastic resistor model for the current from Eq. (3.3) and along

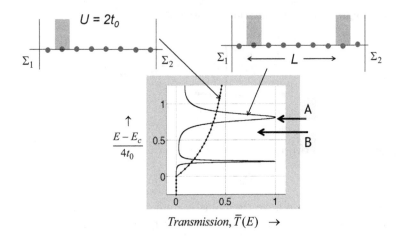

Fig. 19.4 Normalized conductance for a wire with $M = 1$ with one scatterer, and with two scatterers. For MATLAB script, see Appendix H.1.2.

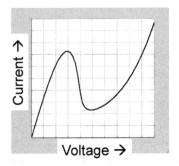

with the conductance function from NEGF

$$G(E) \equiv \frac{q^2}{h}\bar{T}(E) = \frac{q^2}{h}\text{Trace}\left[\mathbf{\Gamma_1 G^R \Gamma_2 G^A}\right]$$

to model devices like this, but it is important to include the effect of the applied electric field on the $\mathbf{H}$ as mentioned earlier (see Fig. 17.6). In this book we will focus more on low bias response for which this aspect can be ignored.

Consider for example a resistor with scatterers distributed randomly throughout the channel. If we were to use the quantum formalism to calculate the conductance function for a single-moded wire with random scatterers we would find that once the classical transmission $M\lambda/L$ drops below one, the quantum conductance is extremely low except for occasional peaks

at specific energies (Fig. 19.5). The result marked semiclassical is obtained
by calculating T for a single scatterer and then increasing the scatterer
contribution by a factor of six:

$$R_6 = \frac{h}{q^2 M} \left(6 \underbrace{\frac{1-T}{T}}_{\text{scatterer}} + \underbrace{1}_{\text{interface}} \right) = \frac{h}{q^2 M} \frac{6 - 5T}{T}.$$

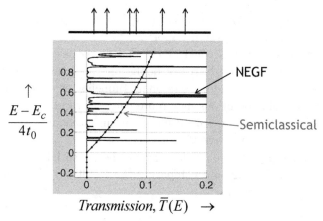

Fig. 19.5 Normalized conductance for a wire with $M = 1$ with six scatterers. For
MATLAB script, see Appendix H.1.3.

Comparing the classical and quantum results suggests that such conductors would generally show very high resistances well in excess of Ohm's
law, with occasional wild fluctuations. In a multi-moded wire too quantum calculations show the same behavior once the classical transmission
$M\lambda/L$ drops below one. Such conductors are often referred to as being in
the regime of *strong localization*. Interestingly, even when $M\lambda/L$ is well in
excess of one, the quantum conductance is a little ($\sim$ approximately one)
less than the classical value and this is often referred to as the regime of
weak localization.

However, localization effects like these are usually seen experimentally
only at low temperatures. At room temperature there is seldom any evidence of deviation from Ohm's law. Consider for instance a copper wire
with a cross-section of 10 nm × 10 nm which should contain approximately
1000 atoms and hence should have $M \sim 1000$ (see discussion at end of
Chapter 4). Assuming a mean free path of 40 nm this suggests that a

copper wire any longer than $M\lambda \sim 40\ \mu$m should exhibit strange non-ohmic behavior, for which there is no experimental evidence. Why?

The answer is that localization effects arise from quantum interference and will be observed only if the entire conductor is **phase-coherent**. A copper wire 40 μm long is not phase coherent, certainly not at room temperature. Conceptually we can think of the real conductor as a series of individual coherent conductors, each of length equal to the phase coherence length L_P and whether we see localization effects will depend not on $M\lambda/L$, but on $M\lambda/L_P$.

The bottom line is that to describe real world experiments especially at room temperature it is often important to include a certain degree of dephasing processes as described at the end of the last chapter. Unless we include an appropriate degree of dephasing our quantum models will show interference effects leading to resonant tunneling or strong localization which under certain conditions may represent real world experiments, but not always. Just because we are using quantum mechanics, the answer is not automatically more "correct".

This can be appreciated by looking at the potential variation along the channel using NEGF and comparing the results to our semiclassical discussion from Chapter 10.

19.3 Potential Drop Across Scatterer(s)

Related video lecture available at course website, Unit 2: L2.9.

In Chapter 10 we discussed the spatial variation of the occupation factor which translates to a variation of the electrochemical potential for low bias. A conductor with one scatterer in it (Fig. 19.6), can be viewed (see Fig. 10.6) as a normalized interface resistance of one in series with a normalized scatterer resistance of $(1 - T)/T$, which can be written as

$$(\text{Normalized})\ R_{\text{scatterer}} = \left(\frac{Ua}{\hbar v}\right)^2 \tag{19.7}$$

using Eq. (19.5). The semiclassical potential profile in Fig. 19.6 is then obtained by noting that since the current is the same everywhere, each section shows a potential drop proportional to its resistance.

The quantum profile is obtained using an NEGF model to calculate the effective occupation factor throughout the channel by looking at the ratio of the diagonal elements of $\mathbf{G}^n$ and $\mathbf{A}$ which are the quantum versions of the electron density and density of states respectively:

$$f(j) = \frac{G^n(j,j)}{A(j,j)}. \tag{19.8}$$

For low bias, this quantity translates linearly into a local electrochemical potential as noted in Chapter 2 (see Eq. (2.11)). If we choose $f = 0$ at one contact, $f = 1$ at another contact corresponding to qV, then the $f(j)$ obtained from Eq. (19.8) is simply translated into an electrochemical potential μ at that point:

$$\mu(j) = qV f(j). \tag{19.9}$$

The occupation $f(j)$ shows oscillations due to quantum interference making it hard to see the potential drop across the scatterer (see solid black line marked NEGF).

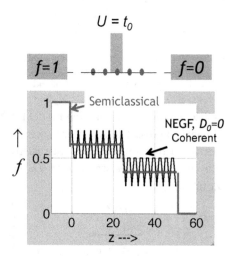

Fig. 19.6   Potential drop across a scatterer calculated from the quantum formalism: Coherent NEGF calculation at $E = t_0$, compared to semiclassical result based on Chapter 10, Part A. For MATLAB script, see Appendix H.1.4.

Experimentalists have measured profiles such as these using scanning probe microscopy (SPM) and typically at room temperature the quantum oscillations are not seen, because of the dephasing processes that are inevitably present at room temperature. This is another example of the need to include dephasing in order to model real world experiments especially at room temperature.

Indeed if we include pure phase relaxation processes (Eq. (18.37)) in the NEGF model we obtain a clean profile looking a lot like what we would expect from a semiclassical picture (see Fig. 19.7a).

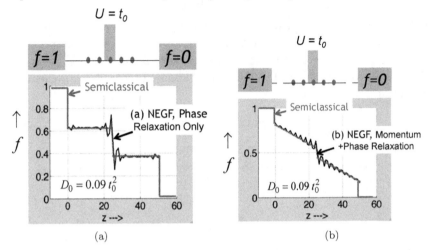

(a)   (b)

Fig. 19.7   Potential drop for the structure in Fig. 19.6 calculated from the NEGF method at $E = t_0$ with dephasing, (a) Phase-relaxation only, Eq. (18.37), (b) Phase and momentum relaxation, Eq. (18.38). For MATLAB script, see Appendix H.1.4.

Interestingly, if we use a momentum relaxing model for Σ_0 (Eq. (18.38)), the potential drops linearly across the structure (see Fig. 19.7b), exactly what we would expect for a distributed classical resistor. The resistance per lattice site for this distributed resistor due to D_0 can be obtained by replacing U^2 with D_0 in Eq. (19.7):

$$\text{(Normalized)} \quad R = \underbrace{\left(\frac{a}{\hbar v}\right)^2 D_0}_{\substack{\text{Resistance} \\ \text{per lattice site}}} \times \underbrace{\frac{L}{a}}_{\substack{\text{\# of lattice} \\ \text{sites}}} .$$

Another interesting example is that of the two quantum resistors in series that we started with. We noted then that at energies corresponding to points A and B in Fig. 19.4 we have constructive and destructive interference respectively. This shows up clearly in the potential profile for coherent transport with $D_0 = 0$ (see Fig. 19.8). At $E = 0.6t_0$ corresponding to destructive interference, the profile looks like what we might expect for a very large resistor showing a large drop in potential around it along with some sharp spikes superposed on it. At $E = 0.81t_0$ corresponding to

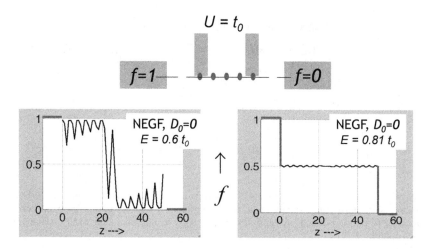

Fig. 19.8 Potential drop across two scatterers in series calculated from the NEGF method without dephasing at two energies, $E = 0.81t_0$ and $E = 0.6t_0$ corresponding to points marked "A" and "B" respectively in Fig. 19.4. For MATLAB script, see Appendix H.1.5.

constructive interference, the profile looks like what we expect for a ballistic conductor with all the drop occurring at the two contacts and none across the scatterers.

Clearly at $E = 0.81t_0$ the answer to the title question of this chapter is yes, two scatterers can offer less resistance than one. And this strange result is made possible by quantum interference. And once we introduce sufficient phase relaxation into the model using a non-zero D_0, the profile at both energies look much the same like any semiclassical resistor (Fig. 19.9).

Before we move on, let me note that although it is straightforward to include dephasing into toy calculations like this, for large devices described by large matrices, it can be numerically challenging. This is because with coherent NEGF ($D_0 = 0$) or with the momentum relaxing model (Eq. (18.38)), it is often adequate to calculate just the diagonal elements of the Green's functions using efficient algorithms. But for pure phase relaxation (Eq. (18.37)), it is necessary to calculate the full Green's function increasing both computational and memory burdens significantly.

So a natural question to ask is whether instead of including dephasing through Σ_0 we could include the potential U_R in the Hamiltonian itself and then average our quantity of interest over a number of random realizations of U_R. Would these results be equivalent?

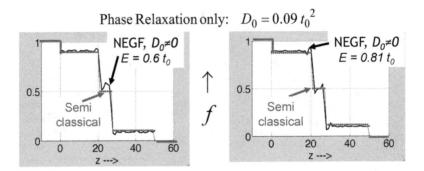

Fig. 19.9   Potential drop across two scatterers in series calculated from the NEGF method with pure phase relxation at two energies, $E = 0.81t_0$ and $E = 0.6t_0$ corresponding to points marked "A" and "B" respectively in Fig. 19.4. For MATLAB script, see Appendix H.1.5.

For short conductors like the one shown in Fig. 19.4, this seems to be true, but for long conductors like the one in Fig. 19.5 this may not be true. With a conductor in the regime of strong localization (Fig. 19.5) it is hard to see how averaging the coherent quantum result over many configurations can lead to the semiclassical result.

NEGF with dephasing does not just average over many configurations, it also averages over different sections of the same configuration and that is why it is able to capture the semiclassical result which often describes real world experiments at room temperature quite well.

But could NEGF capture the localization effects observed at low temperatures through a proper choice of Σ_0? We believe so, but it would involve going beyond the simple dephasing models (technically known as the self-consistent Born approximation) for evaluating Σ_0 described in Section 18.4.

PART 2

More on NEGF

Chapter 20

Quantum of Conductance

Related video lecture available at course website, Unit 3: L3.1.

As I mentioned, our primary objective in Chapters 19–23 is to help the reader get familiar with the NEGF model through "do it yourself" examples of increasing complexity. The last chapter used 1D examples. In this chapter we look at 2D examples which illustrate one of the key results of mesoscopic physics, namely the observation of conductances that are an integer multiple of the conductance quantum q^2/h.

20.1 2D Conductor as 1D Conductors in Parallel

Related video lecture available at course website, Unit 3: L 3.2.

Among the seminal experiments from the 1980s that gave birth to mesoscopic physics was the observation that the conductance of a ballistic 2D conductor went down in integer multiples of $2q^2/h$ as the width of the narrow region was decreased.

To understand this class of experiments we need a 2D model (Fig. 20.1). As with 1D, two inputs are required: the Hamiltonian **H** and the contact self-energy matrices **Σ**. Once we have these, the rest is standard.

For **H**, we use the 2D Hamiltonian from Fig. 17.7 for conductors described by parabolic $E(\mathbf{k})$ relations. As we discussed earlier for a uniform wire the dispersion relation is given by

$$E(k_x, k_y) \;=\; \varepsilon \;+\; 2t\cos\left(k_x a\right) \;+\; 2t\cos\left(k_y a\right) \qquad (20.1a)$$

which can approximate a parabolic dispersion

$$E \;=\; E_c \;+\; \frac{\hbar^2 k^2}{2m} \qquad (20.1b)$$

by choosing

$$E_c = \varepsilon + 4t \qquad (20.2a)$$

and

$$-t \equiv t_0 \equiv \frac{\hbar^2}{2ma^2}. \qquad (20.2b)$$

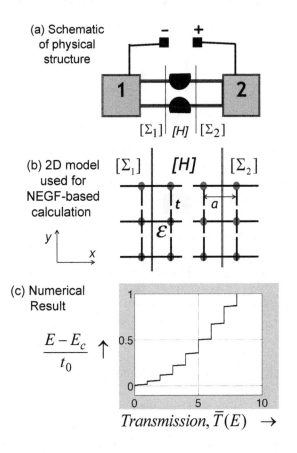

Fig. 20.1 (a) Schematic of structure for measuring the conductance of a short constriction created in a two dimensional conductor. (b) 2D model used for NEGF-based calculation and discussion. (c) Numerically computed transmission shows steps as a function of energy. For MATLAB script, see Appendix H.2.1.

Once again what needs discussion are the **self-energy matrices, Σ**, but before we get into it let us look at the transmission function

$$\bar{T}(E) = \text{Trace}\left[\mathbf{\Gamma}_1 \mathbf{G}^R \mathbf{\Gamma}_2 \mathbf{G}^A\right] \quad \text{(same as Eq. (19.3))}$$

obtained directly from the numerical model (Fig. 20.1), which shows steps at specific energies. How can we understand this?

The elementary explanation from Section 6.4 is that for a ballistic conductor the transmission function is just the number of modes $M(E)$ which equals the number of half de Broglie wavelengths that fits into the width W of the conductor (floor(x) denotes the highest integer less than x)

$$M = \text{floor}\left(\frac{2W}{h/p}\right) = \text{floor}\left(\frac{2W}{h}\sqrt{2m(E - E_c)}\right)$$

where we have used the parabolic relation $E - E_c = p^2/2m$. To compare with our numerical results we should use the cosine dispersion relation.

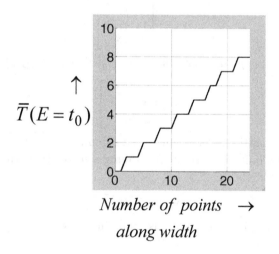

Experimentally what is measured at low temperatures is $M(E = \mu_0)$ and the steps are observed as the width is changed as first reported in van Wees *et al.* (1988) and Wharam *et al.* (1988). To compare with experimental plots, one could take a fixed energy $E = t_0$ and plot the transmission as a function of the number of points along the width to get something like this.

Why does our numerical model show these steps? One way to see this is to note that our 2D model can be visualized as a linear 1D chain as shown in the following figure where the individual elements α of the chain represent a column. For example if there are three sites to each column,

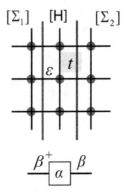

we would have

$$\boldsymbol{\alpha} = \begin{bmatrix} \varepsilon & t & 0 \\ t & \varepsilon & t \\ 0 & t & \varepsilon \end{bmatrix} \qquad (20.3\text{a})$$

while the coupling $\boldsymbol{\beta}$ from one column to the next is diagonal:

$$\boldsymbol{\beta} = \begin{bmatrix} t & 0 & 0 \\ 0 & t & 0 \\ 0 & 0 & t \end{bmatrix}. \qquad (20.3\text{b})$$

Note that the matrix $\boldsymbol{\alpha}$ describing each column has off-diagonal elements t, but we can eliminate these by performing a ***basis transformation to diagonalize it:***

$$\tilde{\boldsymbol{\alpha}} = \mathbf{V}^{\dagger} \boldsymbol{\alpha} \mathbf{V} \qquad (20.3\text{c})$$

where $\mathbf{V}$ is a matrix whose columns represent the eigenvectors of $\boldsymbol{\alpha}$.

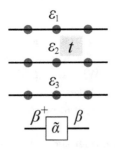

Since the matrix $\boldsymbol{\beta}$ is essentially an identity matrix it is unaffected by the basis transformation, so that in this transformed basis we can visualize the 2D conductors as a set of independent 1D conductors, each of which has a different diagonal element

$$\varepsilon_1, \ \varepsilon_2, \ \varepsilon_3$$

equal to the eigenvalues of α. Each of these 1D conductors has a transmission of one in the energy range $(t_0 \equiv |t|)$

$$\varepsilon_n - 2\,t_0 \ < \ E \ < \ \varepsilon_n + 2\,t_0$$

as sketched below. Adding all the individual transmissions we obtain the transmission showing up-steps in the lower part and down-steps in the upper part.

Usually when modeling n-type conductors we use the lower part of the band as shown in Fig. 20.1, and so we see only the up-steps occurring at

$$\varepsilon_n - 2\,t_0$$

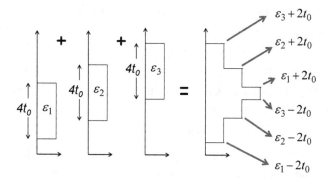

Now the ε_n's are the eigenvalues of α (see Eq. (20.3a)) which are given by

$$\varepsilon_n = \varepsilon - 2t_0 \cos\left(k_n a\right), \quad \text{with} \quad k_n a = \frac{n\pi}{N+1} \tag{20.4}$$

where N is the number of points along the width which determines the size of α. This result is not obvious, but can be shown analytically or checked easily using MATLAB.

Using Eqs. (20.2) and (20.4) we can write the location of the steps as

$$\varepsilon_n - 2t_0 \ = \ E_c + 2t_0\left(1 - \cos\left(\frac{n\pi}{N+1}\right)\right)$$

which matches the numerical result obtained with $N = 25$ very well as shown.

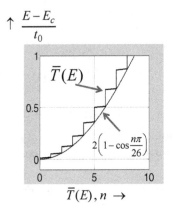

$$\overline{T}(E), n \rightarrow$$

20.1.1   *Modes or subbands*

The approach we just described of viewing a 2D (or 3D) conductor as a set of 1D conductors in parallel is a very powerful one. Each of these 1D conductors is called a mode (or subband) and has a dispersion relation

$$E_n(k_x) = \varepsilon_n - 2t_0 \cos(k_x a)$$

as shown below. These are often called the subband dispersion relations obtained from the general dispersion relation in Eq. (20.1a) by requiring k_y to take on quantized values given by

$$k_y a = \frac{n\pi}{N+1}$$

where each integer n gives rise to one subband as shown. If we draw a horizontal line at any energy E, then the number of dispersion relations it

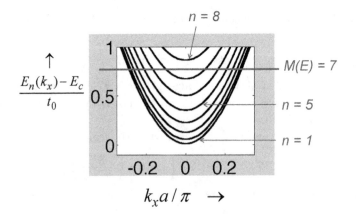

crosses is equal to twice the number of modes $M(E)$ at that energy, since each mode gives rise to two crossings, one for a state with positive velocity, and one for a state with negative velocity.

20.2   Contact Self-Energy for 2D Conductors

Related video lecture available at course website, Unit 3: L3.3.

Let us now address the question we put off, namely how do we write the self-energy matrices for the contacts. Ideally the contact regions allow electrons to exit without any reflection and with this in mind, a simple way to evaluate Σ is to assume the contacts to be just uniform extensions of the channel region and that is what we will do here.

20.2.1   *Method of basis transformation*

The viewpoint we just discussed in Section 20.1 allows us to picture a 2D conductor as a set of decoupled 1D conductors, by converting from the usual lattice basis to an abstract mode basis through a basis transformation:

$$\underbrace{\tilde{\mathbf{X}}}_{\text{Mode Basis}} = \mathbf{V}^{\dagger} \underbrace{\mathbf{X}}_{\text{Lattice Basis}} \mathbf{V} \qquad (20.5a)$$

$\mathbf{X}$ being any matrix in the regular lattice basis. A unitary transformation like this can be reversed by transforming back:

$$\underbrace{\mathbf{X}}_{\text{Lattice Basis}} = \mathbf{V} \underbrace{\tilde{\mathbf{X}}}_{\text{Mode Basis}} \mathbf{V}^{\dagger} \qquad (20.5b)$$

In our present problem we can easily write down the self-energy in the mode basis for each independent 1D wire and then connect them together.

For example if each wire consisted of just one site along x, then each wire would have a self-energy of $t \exp(ika)$, with the appropriate ka for that wire at a given energy E. For mode n we have

$$E = \varepsilon_n - 2t_0 \cos(k_n a)$$

so that overall we could write

$$\tilde{\Sigma}_1 = \begin{bmatrix} te^{ik_1 a} & 0 & 0 & \cdots \\ 0 & te^{ik_2 a} & 0 & \cdots \\ 0 & 0 & te^{ik_3 a} & \cdots \\ \vdots & \vdots & \vdots & \ddots \end{bmatrix}$$

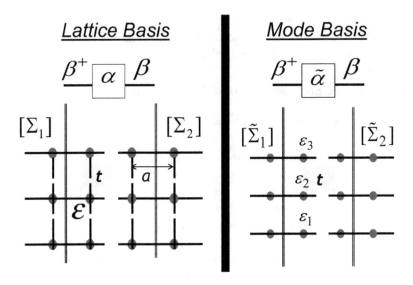

Fig. 20.2   A 2D conductor can be pictured as a set of decoupled 1D conductors through a basis transformation.

and then transform it back to the lattice basis as indicated in Eq. (20.5b):

$$\boldsymbol{\Sigma}_1 \;=\; \mathbf{V}\,\tilde{\boldsymbol{\Sigma}}_1\,\mathbf{V}^\dagger.$$

20.2.2   General method

Related video lecture available at course website, Unit 3: L3.4.

The method of basis transformation is based on a physical picture that is very powerful and appealing. However, I believe it cannot always be used at least not as straightforwardly, since in general it may not be possible to diagonalize both $\boldsymbol{\alpha}$ and $\boldsymbol{\beta}$ simultaneously.

For the square lattice $\boldsymbol{\beta} = t\mathbf{I}$ (Eq. (20.3b)) making it *"immune"* to basis transformations, since the identity matrix remains an identity matrix in all bases. But in general this may not be so. The graphene lattice from Fig. 17.9 pictured below is a good example. How do we write $\boldsymbol{\Sigma}$ in such cases?

Any conductor with a uniform cross-section can be visualized as a linear 1-D chain of "atoms" each having an on-site matrix Hamiltonian $\boldsymbol{\alpha}$ coupled to the next "atom" by a matrix $\boldsymbol{\beta}$. Each of these matrices is of size $(n \times n)$, n being the number of basis functions describing each unit.

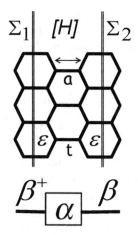

The **self-energy matrix** is zero except for the last $(n \times n)$ block at the surface

$$\Sigma_2(E) = \begin{bmatrix} \ddots & \vdots & \vdots \\ \cdots & 0 & 0 \\ \cdots & 0 & \beta \mathbf{g}_2 \beta^\dagger \end{bmatrix}.$$

The non-zero block is given by

$$\beta\,\mathbf{g}_2\,\beta^\dagger \qquad\qquad (20.6a)$$

where $\mathbf{g}_2$ is called the surface Green's function for contact 2, and is obtained by iteratively solving the equation:

$$\mathbf{g}_2^{-1} = (E + i0^+)\mathbf{I} - \boldsymbol{\alpha} - \boldsymbol{\beta}^\dagger \mathbf{g}_2 \boldsymbol{\beta} \qquad\qquad (20.6b)$$

for $\mathbf{g}_2$, where 0^+ represents a positive infinitesimal. Equation (20.6) is of course not meant to be obvious, but we have relegated the derivation to Appendix G. We will not go into the significance of the infinitesimal $i0^+$ (see for example Datta (1995), Chapter 3 or Datta (2005), Chapter 8).

For the moment let me just note that for a 1D conductor with $\boldsymbol{\alpha} = \varepsilon$ and $\boldsymbol{\beta} = t$ Eq. (20.2) reduces to an ordinary quadratic equation:

$$g_2\left(E + i0^+ - \varepsilon - t^2 g_2\right) = 1$$

whose solution gives two possible solutions $t\exp\left(\pm ika\right)$ for the self-energy, and the one we want is that with the *negative imaginary part*, for which *the corresponding broadening* $\mathbf{\Gamma}$ *is positive*. More generally, we have a matrix quadratic equation (Eq. (20.6b)) and the infinitesimal $i0^+$ ensures that a numerical iterative solution converges on the solution for which $\mathbf{\Gamma}$ has all positive eigenvalues.

20.2.3   Graphene: ballistic conductance

Related video lecture available at course website, Unit 3: L3.5.

As an example we have shown in Fig. 20.3 the transmission $\bar{T}(E)$ calculated numerically for two common orientations of graphene, the so-called zigzag and armchair configurations with dimensions chosen so as to have roughly equal widths. Since these are ballistic conductors, the transmission is equal to the number of modes $M(E)$ and can be approximately described by the number of wavelengths that fit into the widths. The actual energy dependence is different from that obtained for the square lattice (see Eq. (20.3)) because of the linear $E(\mathbf{k})$ relation: $E = \hbar\nu_0 k = \nu_0 p$:

$$M = \text{floor}\left(\frac{2W}{h/p}\right) = \text{floor}\left(\frac{2W}{h}\frac{E}{\nu_0}\right). \qquad (20.7)$$

This applies equally to any orientation of graphene. Both the orientations shown have the same overall slope, but the details are quite different. For example, at $E = 0$, the armchair is non-conducting with $M = 0$ while the zigzag is conducting with non-zero M.

For large dimensions the steps are close together in energy (compared to kT) and both appear to be semi-metallic. But for small dimensions the steps are much larger than kT. The zigzag now shows zero transmission $\bar{T}(E) = 0$ at $E = 0$ ("semiconducting") while the armchair shows non-zero conductance ("metallic"). These are clear observable differences that show up in experiments on samples of small width at low temperatures.

Another interesting observable difference is that between a flat graphene sheet and a cylindrical carbon nanotube (CNT). Mathematically, they are both described by the same Hamiltonian $\mathbf{H}$ but with different boundary conditions. Graphene like most normal conductors requires "hardwall boundary conditions" (HBC) where the lattice ends abruptly at the edges. CNT's on the other hand are among the few real conductors that require "periodic boundary conditions" (PBC) with no edges.

The results for CNT are relatively easy to understand analytically, while those for graphene require a more extensive discussion (see for example Brey and Fertig, 2006). As we mentioned in Chapter 6, PBC is mathematically

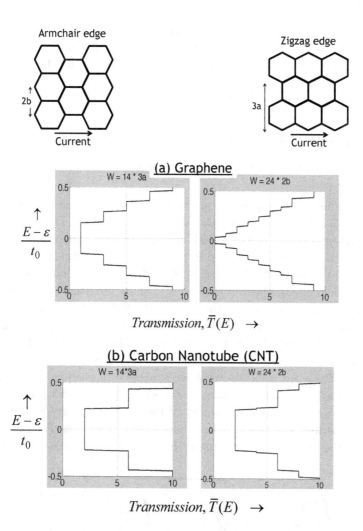

Fig. 20.3 $\bar{T}(E)$ calculated from NEGF-based model for a ballistic (a) graphene sheet with armchair and zigzag edges as shown with roughly equal widths ($24 \times 2b \approx 14 \times 3a$), (b) carbon nanotube (CNT) obtained by rolling up the graphene sheet along the width.

simpler and that is why it is used so extensively for large conductors where it is known experimentally that the exact boundary conditions are not very relevant. But this of course is not true of small conductors and the difference is evident in Fig. 20.3 for small conductors only a few nanometers in width. We will not go into this further. Our objective here is simply to show how easily our quantum transport formalism captures all the known physics.

The power of the numerical method lies in being able to calculate $M(E)$ automatically even before one has "understood" the results. However, one should use numerical calculations not as a substitute for understanding, but as an aid to understanding.

20.3 Quantum Hall Effect

Related video lecture available at course website, Unit 3: L3.6.

The Hall effect (Chapter 11) provides another good example for a two-dimensional application of the quantum transport model. The basic structure involves a long conductor with side probes designed to measure the transverse Hall voltage developed in the presence of a magnetic field.

We use the same 2D Hamiltonian from Fig. 20.1 but now including a magnetic field as explained in Section 17.4.3. As discussed in Chapter 13, the Hall resistance is given by the ratio of the Hall voltage to the current. In a theoretical model we could calculate the Hall voltage in one of two ways. We could attach a voltage probe to each side and use Büttiker's multiterminal method to find the potentials they float to.

Alternatively we could do what we explained in Section 19.3, namely calculate the fractional occupation of the states at any point j by looking at the ratio of the diagonal element of the electron density $\mathbf{G}^n$ and the density of states $\mathbf{A}$ and use the low bias Taylor expansion (Eq. (2.11)) to translate the occupation factor profile into a potential profile.

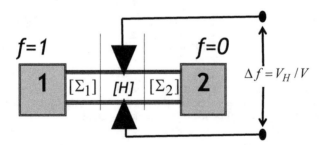

Figure 20.4 shows the calculated Hall resistance (normalized to the quantum of resistance h/q^2) as a function of the magnetic field. The striking result is of course the occurrence of plateaus at high fields known as the quantum Hall effect (von Klitzing *et al.*, 1980). But first let us note

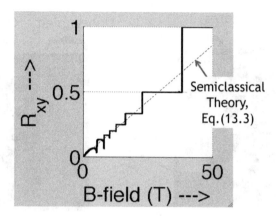

Fig. 20.4 Normalized Hall resistance versus B-field for ballistic channel of width $W = 26a = 65$ nm calculated at an energy $E = t_0$ using a 2D model from Fig. 20.1. For MATLAB script, see Appendix H.2.3.

the low field regime where the calculated result matches the Hall resistance expected from semiclassical theory

$$R_H = \frac{B}{qn} \quad \text{(same as Eq. (11.6)).}$$

The dashed line in Fig. 20.3 is obtained from Eq. (11.6) assuming

$$\frac{N}{LW} = \frac{k^2}{4\pi} \quad \text{(from Eq. (6.17) using } p = \hbar k)$$

and noting that the numerical calculation is carried out at $E = t$, corresponding to $ka = \pi/3$, with $a = 2.5$ nm.

The semiclassical theory naturally misses the high field results which arise from the formation of Landau levels due to quantum effects. These are evident in the numerical plots of the local density of states at high **B**-field (20 T) shown in Fig. 20.5.

Usually the density of states varies relatively gently with position, but in the quantum Hall regime, there is a non-trivial modification of the local density of states which can be plotted from the NEGF method by looking at the diagonal elements of the spectral function $A(j, j; E)$. Figure 20.5 is a grayscale plot of $A(j, j; E)$ with energy E on the horizontal axis and the position j along the width on the vertical axis. The white streaks indicate a high density of states corresponding to the energy of Landau levels, which increase in energy along the edge forming what are called *edge states*.

As we mentioned in Chapter 11, the edge states can be pictured semiclassically in terms of *"skipping orbits"* that effectively isolate oppositely

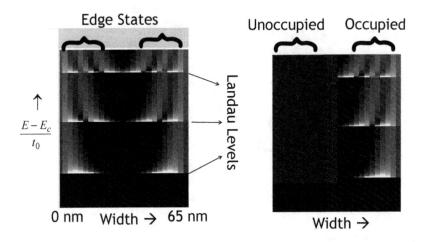

Fig. 20.5   Grayscale plot of local density of states, obtained from the diagonal elements of $\mathbf{A}(E)$ calculated at $B = 20$ T from the NEGF method. Also shown on the right are the diagonal elements of $\mathbf{G}^n(E)$ calculated assuming $f_1 = 1$, $f_2 = 0$. For MATLAB script, see Appendix H.2.4.

moving electrons from each other giving rise to a "*divided highway*" that provides an incredible level of *ballisticity*. This is evident if we plot the electron density from the diagonal elements of $\mathbf{G}^n$ under non-equilibrium conditions assuming $f_1 = 1$, $f_2 = 0$. Only the edge states on one side of the sample are occupied. If we reverse the current flow assuming $f_1 = 0$, $f_2 = 1$, we would find the edge states on the other side of the conductor occupied.

The energies of the Landau levels are given by

$$E_n = \left(n + \frac{1}{2}\right) \hbar\omega_c \tag{20.8}$$

where n is an integer, ω_c being the cyclotron frequency (see Eq. (11.10)). We expect the streaks to be spaced by

$$\hbar\omega_c = \frac{\hbar qB}{m} = \frac{2qBa^2}{\hbar} t_0$$

where we have made use of Eqs. (11.10) and (20.2b). Noting that $B = 20$ T, $a = 2.5$ nm, we expect a spacing of $\sim 0.37\, t_0$ between the streaks in approximate agreement with Fig. 20.5.

Equation (20.8) is a quantum result that comes out of the Schrödinger equation including the vector potential which is part of our numerical model. One can understand it heuristically by noting that semiclassically electrons describe circular orbits in a magnetic field, completing one orbit in a time (see Eq. (11.10))

$$t_c = \frac{2\pi}{\omega_c} = \frac{2\pi p}{q\nu B}$$

so that the circumference of one orbit of radius r_c is given by

$$2\pi r_c = \nu t_c = \frac{2\pi p}{qB}.$$

If we now impose the quantum requirement that the circumference equals an integer number of de Broglie wavelengths h/p, we have

$$\frac{2\pi p}{qB} = \text{integer} \times \frac{h}{p}.$$

Semiclassically an electron can have any energy $E = p^2/2m$. But the need to fit an integer number of wavelengths leads to the condition that

$$p^2 = \text{integer} \times hqB$$

suggesting that the allowed energies should be given by

$$E = \text{integer} \times \frac{hqB}{2m} = \text{integer} \times \frac{\hbar\omega_c}{2}$$

which is not exactly the correct answer (Eq. (20.8)), but close enough for a heuristic argument.

The resulting current equals

$$\frac{q^2}{h} V \times \text{Number of Edge States}$$

while the Hall voltage simply equals the applied voltage since one edge of the sample is in equilibrium with the source and other with the drain.

This leads to a quantized Hall resistance given by

$$\frac{h}{q^2} \times \frac{1}{\text{Number of Edge States}}$$

giving rise to the plateaus of $1/4, 1/3, 1/2, 1$ seen in Fig. 20.4, as the magnetic field raises the Landau levels, changing the number of edge states at an energy $E = t_0$ from 4 to 3 to 2 to 1.

I should mention that the theoretical model does not include the two spins and so gives a resistance that is twice as large as the experimentally observed values which look more like

$$\frac{h}{2q^2} \times \frac{1}{\text{Number of Edge States}}$$

because edge states usually come in pairs, except at high **B**-fields.

Also, we have not talked at all about the fractional quantum Hall effect observed in pure samples at larger **B**-fields with Hall resistances that look like

$$\frac{h}{q^2} \times \frac{1}{\text{a fraction}}.$$

This is a vast and rich area of research on its own beyond the scope of the simple NEGF model discussed here. As it stands it captures only the integer Hall effect though innovative extensions could take it beyond this regime.

Chapter 21

Inelastic Scattering

Related video lecture available at course website, Unit 3: L3.8.

Back in Chapter 17 we used this picture (Fig. 21.1) to summarize our NEGF model in which the channel is described by a Hamiltonian **H** while the self-energies $\boldsymbol{\Sigma}_1$ and $\boldsymbol{\Sigma}_2$ describe the exchange of electrons with the physical contacts. $\boldsymbol{\Sigma}_0$ describes the interactions with the surroundings which can be viewed as additional conceptual "contacts".

Given these inputs, the basic NEGF equations (see Eqs. (18.1)–(18.4)) tell us how to analyze any given structure. Since then we have been looking at various examples illustrating how one writes down **H** and $\boldsymbol{\Sigma}$ and uses the NEGF equations to extract concrete results and investigate the physics. One major simplification we have adopted is in our treatment of the interactions in the channel represented by $\boldsymbol{\Sigma}_0$ which we have either ignored (coherent transport) or treated as an elastic dephasing process described by Eqs. (18.35).

This choice of self-energy functions leads to no exchange of energy with the surroundings, but it has an effect on transport due to the exchange of momentum and "phase". Basically we have been talking about elastic resistors like the ones we started this book with, except that we are now including quantum mechanical effects. One could say that in the last few chapters we have applied the general Non-Equilibrium Green's Function (NEGF) method to an elastic resistor, just as in Part A we applied the general Boltzmann Transport Equation (BTE) to an elastic resistor.

So how do we go beyond elastic resistors? For semiclassical transport, it is clear in principle how to include different types of interaction into the BTE for realistic devices and much progress has been made in this direction. Similarly for quantum transport, the NEGF tells us how to

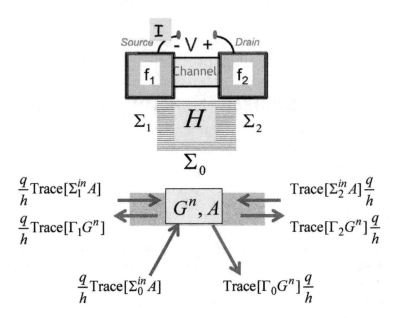

Fig. 21.1 NEGF model: Channel is described by **H** while the effect of contacts is described by $\boldsymbol{\Sigma}_1$ and $\boldsymbol{\Sigma}_2$. Scattering processes are like an abstract contact whose effect is described by $\boldsymbol{\Sigma}_0$.

evaluate the self-energy $\boldsymbol{\Sigma}_0$ for any given microscopic interaction. In this book we have talked only about elastic dephasing which is a small subset of the interactions considered in the classic work on NEGF (see for example, Danielewicz, 1984 or Mahan, 1987).

In practice, however, it remains numerically challenging to go beyond elastic resistors and approximate methods continue to be used widely. Readers interested in the details of device analysis at high bias may find an old article (Datta (2000)) useful. This article has a number of concrete results obtained using MATLAB codes that I had offered to share with anyone who asked me for it. Over the years many have requested these codes from me which makes me think they may be somewhat useful and we plan to have these available on our website for these notes.

I should mention that many devices are rather forgiving when it comes to modeling the physics of inelastic scattering correctly. Devices with energy levels that are equally connected to both contacts (Fig. 12.6b) do not really test the deep physics of inelastic transport and cannot distinguish between a good theory and a bad one. A good test for inelastic scattering models is the device shown in Fig. 12.6a for which the entire terminal current is driven by inelastic processes. Only a fundamentally sound theory like NEGF will predict results that comply with the requirements of the second law.

Earlier in Section 18.4 we discussed how to write these functions for elastic dephasing

$$\left[\Sigma_0^{in}(E)\right]_{ij} = D_{ij} \times \left[G^n(E)\right]_{ij} \tag{21.1a}$$

$$\left[\Gamma_0(E)\right]_{ij} = D_{ij} \times \left[G^n(E) + G^p(E)\right]_{ij} \tag{21.1b}$$

where we have introduced a new symbol $\mathbf{G}^p = \mathbf{A} - \mathbf{G}^n$. As we discussed earlier $\mathbf{A}/2\pi$ and $\mathbf{G}^n/2\pi$ represent matrix versions of density of states and electron density respectively. It then follows that their difference $\mathbf{G}^p/2\pi$ represents the matrix version of the *"hole"* density.

In this chapter my objective is to present the corresponding results for inelastic scattering. But first let me step back and try to explain a key distinction between elastic and inelastic processes. Elastic processes are bidirectional, they proceed equally well in either direction. Inelastic processes proceed more easily when the system loses energy than when it gains energy, that is emission is easier than absorption as we had discussed in Chapter 15, Part A.

This distinction arises because inelastic scatterers like phonons change their state in the process, while elastic scatterers are usually rigid with no internal degrees of freedom. Elastic scatterers with internal degrees of freedom can lead to scattering that is not bidirectional as we argued in Chapter 16, Part A for a collection of non-interacting spins that could function like a Maxwell's demon when it is out-of-equilibrium (Section 16.2). In this chapter we will not consider such unusual scatterers, or indeed any specific scatterer.

21.1 Fermi's Golden Rule

Related video lecture available at course website, Unit 3: L3.7.

All quantum mechanics texts describe the celebrated *Fermi's golden rule* which is widely used to model scattering processes. Let us see how we can obtain this result from our NEGF expressions.

In Section 20.2.2 (and Appendix G) we partitioned the total Hamiltonian $\mathbf{H}_{\text{total}}$ for the structure into a channel Hamiltonian $\mathbf{H}$ and a contact (or reservoir) Hamiltonian $\mathbf{H}_R$ coupled by a matrix τ:

$$\mathbf{H}_{\text{total}} = \begin{bmatrix} \mathbf{H} & \tau^\dagger \\ \tau & \mathbf{H}_R \end{bmatrix}. \tag{21.2}$$

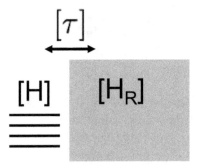

Fig. 21.2   Partitioning into channel $\mathbf{H}$ and contact (or reservoir) $\mathbf{H}_R$ with coupling τ.

We used this partitioning to obtain a general expression for the self-energy function

$$\mathbf{\Sigma} = \tau^\dagger \, \mathbf{g} \, \tau \tag{21.3}$$

in terms of the coupling matrix τ and the Green's function 'g' for the isolated contact

$$\mathbf{g}(E) = \left[(E + i0^+)\mathbf{I} - \mathbf{H}_R \right]^{-1} . \tag{21.4}$$

We can relate the *broadening* $\mathbf{\Gamma} = i[\mathbf{\Sigma} - \mathbf{\Sigma}']$ to the *spectral function* $\mathbf{a} = i[\mathbf{g} - \mathbf{g}']$ using Eq. (21.3)

$$\mathbf{\Gamma} = \tau^\dagger \, \mathbf{a} \, \tau. \tag{21.5}$$

We will now discuss the golden rule and show that it follows from Eq. (21.5), first for elastic scattering and then for inelastic scattering.

21.1.1   *Elastic scattering*

Consider a channel with a static potential $U_s(\mathbf{r})$ that scatters elections from an incident eigenstate $\exp{(+i\mathbf{k} \cdot \mathbf{r})}$ to a final eigenstate $\exp{(+i\mathbf{k}' \cdot \mathbf{r})}$ as shown in Fig. 21.3. The standard golden rule result for the scattering rate is given by

$$S(\mathbf{k}' \leftarrow \mathbf{k}) = \frac{2\pi}{\hbar} |\tau_{\mathbf{k}',\mathbf{k}}|^2 \, \delta(E_k - E_{k'}) \tag{21.6}$$

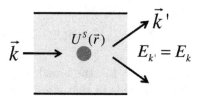

Fig. 21.3   Elastic scattering of electrons by a static potential $U_s(\mathbf{r})$.

where $\tau_{\mathbf{k'},\mathbf{k}}$ is the *"matrix element"* obtained from the scattering potential $U_s(\mathbf{r})$ by performing the integral

$$\tau_{\mathbf{k'},\mathbf{k}} = \int d\mathbf{r} \underbrace{\left(\frac{1}{\sqrt{\Omega}} e^{(+i\mathbf{k'}\cdot\mathbf{r})} \right)^*}_{\text{final state}} U_s(\mathbf{r}) \underbrace{\left(\frac{1}{\sqrt{\Omega}} e^{(+i\mathbf{k}\cdot\mathbf{r})} \right)}_{\text{initial state}}. \tag{21.7}$$

How can we obtain this standard result (Eqs. (21.6) and (21.7)) from the expression we obtained earlier (Eq. (21.5))? The key idea is to visualize the scattering problem depicted in Fig. 21.3 as an abstract channel-contact partitioning problem: the initial state $\mathbf{k}$ is the channel while all the other states $\mathbf{k'}$ form the contact, as shown in Fig. 21.4.

Once we accept this translation from the scattering problem to the partitioning problem, we can write the broadening for the "channel" from Eq. (21.5):

$$\Gamma_{\mathbf{k},\mathbf{k}}(E) = \sum_{\mathbf{k'}} \tau_{\mathbf{k},\mathbf{k'}}^\dagger \, a_{\mathbf{k'},\mathbf{k'}}(E) \, \tau_{\mathbf{k'},\mathbf{k}} \tag{21.8}$$

where we have written out the matrix product explicitly and made use of the fact that the spectral function $\mathbf{a}$ is diagonal. This is because the $\mathbf{k}$-states are eigenstates making $\mathbf{H}$ diagonal, leading to a diagonal Green's function

$$g_{\mathbf{k'},\mathbf{k'}}(E) = \left[E + i0^+ - H_{\mathbf{k'},\mathbf{k'}} \right]^{-1} = \frac{1}{E + i0^+ - E_{k'}}$$

and hence a *diagonal spectral function*:

$$a_{\mathbf{k'},\mathbf{k'}}(E) = i \left(\frac{1}{E + i0^+ - E_k} - \frac{1}{E - i0^+ - E_k} \right) \to 2\pi\delta(E - E_{k'}) \tag{21.9}$$

making use of one definition of the delta function:

$$2\pi\delta(x) = i \left(\frac{1}{x + i0^+} - \frac{1}{x - i0^+} \right).$$

Using Eq. (21.9) in Eq. (21.8), we obtain

$$\Gamma_{\mathbf{k},\mathbf{k}}(E) = \sum_{\mathbf{k}'} \left(\tau_{\mathbf{k}',\mathbf{k}}\right)^* 2\pi\delta(E - E_k)\, \tau_{\mathbf{k}',\mathbf{k}}. \tag{21.10}$$

As we discussed in Section 18.1.3, the broadening of a state is related to the inverse of the time it spends in that state before getting scattered, or the inverse "lifetime" of the state

$$\frac{\hbar}{\text{Lifetime}(\mathbf{k})} = \Gamma_{\mathbf{k},\mathbf{k}}(E_k) = 2\pi \sum_{\mathbf{k}'} |\tau_{\mathbf{k}',\mathbf{k}}|^2\, \delta(E_k - E_{k'})$$

leading to the following expression for the inverse lifetime

$$\frac{1}{\text{Lifetime}(\mathbf{k})} = \frac{2\pi}{\hbar} \sum_{\mathbf{k}'} |\tau_{\mathbf{k}',\mathbf{k}}|^2\, \delta(E_k - E_{k'})$$

which can be written as the sum of all the scattering rates into different states $\mathbf{k}$ given by the golden rule that we stated earlier in Eq. (21.6):

$$\frac{1}{\text{Lifetime}(\mathbf{k})} = \sum_{\mathbf{k}'} S(\mathbf{k}' \leftarrow \mathbf{k}). \tag{21.11}$$

Before moving onto inelastic scattering, let me note that the above discussion applies to the scattering from any initial eigenstate m to any final eigenstate m', which can be viewed as the channel and contact respectively as shown in Fig. 21.4:

$$\frac{1}{\text{Lifetime}(m)} = \frac{2\pi}{\hbar} \sum_{m'} |\tau_{m',m}|^2\, \delta(E_m - E_{m'}) \tag{21.12}$$

which can be written as

$$\frac{1}{\text{Lifetime}(m)} = \sum_{m'} S(m' \leftarrow m) \tag{21.13}$$

where the golden rule scattering rates are given by (cf. Eq. (21.6))

$$S(m' \leftarrow m) = \frac{2\pi}{\hbar} |\tau_{m',m}|^2\, \delta(E_m - E_{m'}) \tag{21.14}$$

with the matrix elements $\tau(m', m)$ calculated using the appropriate eigenfunctions $\phi_m(\mathbf{r})$ for the eigenstates:

$$\tau_{m',m} = \int d\mathbf{r}\ \underbrace{\phi_{m'}^*(\mathbf{r})}_{\text{final state}}\ U_s(\mathbf{r})\ \underbrace{\phi_m(\mathbf{r})}_{\text{initial state}}. \tag{21.15}$$

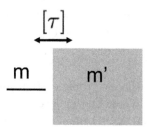

Fig. 21.4   More generally we can visualize the initial state m as the channel and all the other states m' as the contact.

21.1.2   *Inelastic scattering*

A time-dependent scattering potential causes inelastic scattering, that is, scattering with a change in energy.

This can be understood by noting that a sinusoidal scattering potential multiplying an initial wavefunction with energy E_i gives rise to a wave function with two new energies $E_i \pm \hbar\omega$:

$$\underbrace{2U_s(\mathbf{r}) \cos \omega t \times \psi(\mathbf{r}) \exp\left(-\frac{E_i t}{\hbar}\right)}_{\text{Initial state}}$$

$$\rightarrow (U_s\psi)\left[\underbrace{\exp\left(\frac{-(E_i + \hbar\omega)t}{\hbar}\right)}_{\text{absorption}} + \underbrace{\exp\left(\frac{-(E_i - \hbar\omega)t}{\hbar}\right)}_{\text{emission}}\right]$$

$$\bullet \; 2U_s(\vec{r}) \; \cos \omega t$$

This heuristic argument would suggest that the scattering rates for absorption and emission should be equal. However, the correct answer is that they are not equal:

$$S(m' \leftarrow m) = \frac{2\pi}{\hbar}|\tau_{m',m}|^2 \; N_\omega \; \delta(E_m - E_{m'} + \hbar\omega) \; \text{ABSORPTION} \quad (21.16\text{a})$$

$$S(m' \leftarrow m) = \frac{2\pi}{\hbar}|\tau_{m',m}|^2 \; (N_\omega + 1) \; \delta(E_m - E_{m'} - \hbar\omega) \; \text{EMISSION} \quad (21.16\text{b})$$

where N_ω is the number of "*phonons*", while the scattering potential U_s due to one phonon is used to calculate the matrix element $\tau_{m,m}$. Absorption rate is $\sim N_\omega$ while emission rate is $\sim (N_\omega + 1)$.

Now let us see how this result for inelastic scattering can be understood from our earlier result for elastic scattering, namely Eq. (21.14)

$$S(m' \leftarrow m) = \frac{2\pi}{\hbar} |\tau_{m',m}|^2 \, \delta(E_m - E_{m'}).$$

The key idea is to view the electronic system and the "phonon" system as one composite system whose energy levels are sketched in Fig. 21.5 for states with $N_\omega - 1, N_\omega$ and $N_\omega + 1$ phonons. What we normally view as an inelastic absorption process

$$m' \leftarrow m \quad \text{with} \quad E_{m'} = E_m + \hbar\omega$$

becomes an elastic process in the composite picture

$$(m', N_\omega - 1) \leftarrow (m, N_\omega) \quad \text{with} \quad E_{m'} + (N_\omega - 1) \times \hbar\omega = E_m + N_\omega \times \hbar\omega$$

so that the argument of the delta function becomes the same as that in Eq. (21.16a).

$$\big(E_m + N_\omega \times \hbar\omega\big) - \big(E_m + (N_\omega - 1) \times \hbar\omega\big) = E_m - E_{m'} + \hbar\omega.$$

In the composite picture, the scattering rates for both emission and absorption are proportional to the *larger* of the number of phonons in the initial and final states. Absorption takes the system from N_ω to $N_\omega - 1$ and so the scattering rate $\sim N_\omega$, while emission takes the system from N_ω to $N_\omega + 1$ and so the scattering rate $\sim N_\omega + 1$.

21.2 Self-energy Functions

Now that we have seen how our basic result for broadening (Eq. (21.5)) leads to the standard expressions for the golden rule treatment of inelastic scattering, let us get back to the main objective of this chapter, namely to show how the expressions for in-scattering Σ_0^{in} and broadening Γ_0 functions from Section 18.4 for elastic dephasing (Eqs. (21.1a) and (21.1b)) are generalized to include inelastic processes.

The corresponding results for inelastic scattering for an exchange of energy $\hbar\omega$ (> 0 for emission, < 0 for absorption) are given by

$$\big[\Sigma_0^{in}(E)\big]_{ij} = D_{ij}(\hbar\omega) \times \big[G^n(E + \hbar\omega)\big]_{ij}, \tag{21.17a}$$

$$\big[\Gamma_0(E)\big]_{ij} = D_{ij}(\hbar\omega) \times \big[G^n(E + \hbar\omega) + G^p(E - \hbar\omega)\big]_{ij}. \tag{21.17b}$$

The inscattering function in Eq. (21.17a) looks reasonable since we expect the in-scattering at energy E to arise from the electron density (G^n) at energy $E + \hbar\omega$. Emission processes correspond to $\hbar\omega > 0$ while absorption

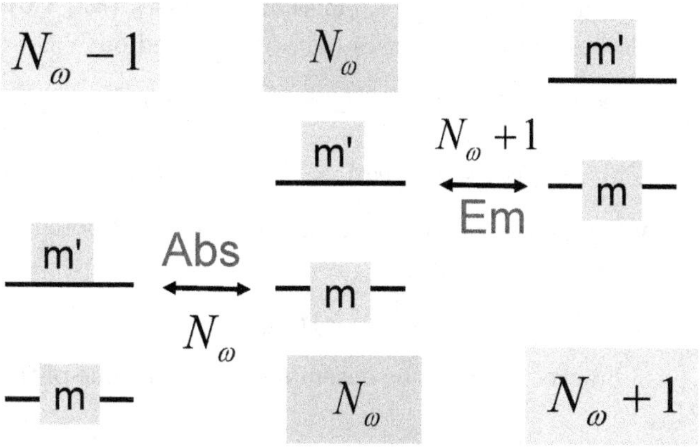

$N_\omega - 1$ N_ω m'

m' $N_\omega + 1$ — m —
Em

m' Abs — m —
N_ω

— m — N_ω $N_\omega + 1$

Fig. 21.5 Energy levels of composite electron-phonon system. Note that emission (Em) and absorption (Abs) involve different final states for the phonon system as shown on the left and their strength depends on the number of phonons in the initial or the final state, whichever is larger. This makes emission processes stronger ($\sim N_\omega + 1$) than absorption processes ($\sim N_\omega$).

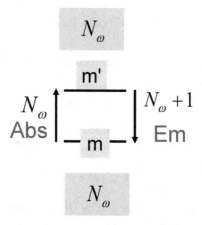

Fig. 21.6 When phonon states are not indicated explicitly, the picture in Fig. 21.5 appears as shown above, as if absorption and emission processes connect the same initial and final states, with upward transitions ($\sim N_\omega$) weaker than downward transitions ($\sim N_\omega + 1$).

processes correspond to $\hbar\omega < 0$. The broadening function in Eq. (21.17b) can be understood as the sum of the inscattering function in Eq. (21.17a) and an out scattering function Σ_0^{out} which is proportional to the number empty states or "holes" at $E - \hbar\omega$.

As we discussed in Section 21.1, emission processes ($\hbar\omega > 0$) depends on $N_\omega + 1$ while absorption processes ($\hbar\omega < 0$) depend on N_ω:

$$D_{ij}(\hbar\omega) = \overline{D}_{ij}(\hbar\omega) \times (N_\omega + 1) , \quad \hbar\omega > 0 \tag{21.18a}$$

$$D_{ij}(\hbar\omega) = \overline{D}_{ij}(\hbar\omega) \times N_\omega , \quad \hbar\omega < 0 \tag{21.18b}$$

where the quantity $\overline{D}_{ij}$ is calculated from the scattering potential U_s due to a single phonon:

$$\overline{D}_{ij} = \langle U_s(i) U_s(j) \rangle \tag{21.19}$$

where the $\langle \cdots \rangle$ brackets denote the ensemble-averaged value of the random potential U_s.

The indices i and j above refer to locations in real space. More generally if we use arbitrary basis functions, the results in Eqs. (21.17a) and (21.17b) have to be generalized to

$$\left[\Sigma_0^{in}(E)\right]_{ij} = \sum_{m,n} \overline{D}_{im;jn}(\hbar\omega) \times \left[G^n(E + \hbar\omega)\right]_{mn} \tag{21.20a}$$

$$\left[\Gamma_0(E)\right]_{ij} = \sum_{m,n} \overline{D}_{im;jn}(\hbar\omega) \times \left[G^n(E + \hbar\omega) + G^p(E - \hbar\omega)\right]_{mn} \tag{21.20b}$$

where $\overline{D}$ is a fourth order tensor obtained from the product of the matrix elements

$$\overline{D}_{i,m;j,n} = \tau_{i,m}\tau_{j,n}^*. \tag{21.21}$$

The matrix elements of τ are obtained as discussed in the last section (Eq. (21.15)), but using the scattering potential U_s due to a single phonon:

$$\tau_{i,m} = \int d\mathbf{r} \, \phi_i^*(\mathbf{r}) \, U_s(\mathbf{r}) \, \phi_m(\mathbf{r}) \tag{21.22}$$

Eqs. (21.20a) and (21.20b) reduce to the simpler versions in Eqs. (21.17a) and (21.17b) if $\tau_{i,m} = \tau_{i,i}\delta_{im}$ so that $\overline{D}_{im;jn}$ can be replaced by just $\overline{D}_{ij}$ where

$$\overline{D}_{i,j} \to \overline{D}_{i,i;j,j} = \tau_{i,i}\tau_{j,j}^*.$$

Chapter 22

Does NEGF Include "Everything?"

Related video lecture available at course website, Unit 3: L3.9.

In the last chapter we have seen how to include inelastic interactions through the self-energy function Σ_0 in the simplest approximation, technically known as the *self-consistent Born approximation*. This is just a small subset of the interactions considered in the classic work on NEGF which provide clear prescriptions for including any microscopic interaction to any degree of approximation (see for example, Danielewicz, 1984 or Mahan, 1987). In practice, however, it remains numerically challenging to include interactions even in the lowest approximations. But practical issues apart, can the NEGF method model *"everything"*, at least in principle?

The formal NEGF method developed in the 1960s was based on many-body perturbation theory (MBPT) which provided clear prescriptions for evaluating the self-energy functions Σ and Σ^{in} for a given microscopic interaction up to any order in perturbation theory. It may seem that using MBPT we can in principle include everything. However, I believe this is not quite true since it is basically a perturbation theory which in a broad sense amounts to evaluating a quantity like $(1 - x)^{-1}$ by summing a series like $1 + x + x^2 + x^3 + \cdots$, which works very well if x is much less than one. But if x happens to exceed one, it does not work and one needs non-perturbative methods, or perhaps a different perturbation parameter.

This is one of the reasons I prefer to decouple the NEGF equations (Eqs. (18.1) through (18.4)) from the MBPT-based methods used to evaluate the self-energy functions. The latter may well evolve and get supplemented as people find better approximations that capture the physics in specific situations.

With equilibrium problems, for example, density functional theory (DFT)-based techniques have proven to be very successful and are often used in quantum chemistry in place of MBPT. I believe one should be cautious about expecting the same success with non-equilibrium problems where a far greater spectrum of many body states are made accessible and can be manipulated through a judicious choice of contacts, but it is quite likely that people will find insightful approaches that capture the essential physics in specific problems.

Like the BTE for semiclassical transport, NEGF-based methods in their simplest form, seem to provide a good description of problems where electron-electron interactions can be treated within a *mean field theory* based on the widely used picture of quasi-independent electrons moving in a self-consistent potential U due to the other electrons (Section 17.2).

As we saw in Chapter 8, for low bias calculations one needs to consider only the equilibrium potential which is already included in the semi-empirical tight-binding (TB) parameters used to construct our Hamiltonian **H**. For real devices operating at high bias, the change in the potential due to any changes in the electron occupation in the channel are routinely included using the Poisson equation which is the simplest approximation to the very difficult problem of electron-electron interactions and there have been extensive discussions of how the self-consistent field (SCF) can be corrected to obtain better agreement with experimental results.

However, there are examples where the self-consistent field approach itself seems to fail and some of the most intriguing properties arise from a failure of this simple picture. The purpose of this chapter is to alert the reader that a straightforward application of NEGF may well miss these important experimentally observable effects. Future challenges and opportunities may well involve effects of this type, requiring insightful choices for Σ and Σ^{in} if we wish to use the NEGF method.

22.1   Coulomb Blockade

Let us consider the simplest resistor that will show this effect, one that is only slightly more complicated than the one-level resistor we started with (Fig. 3.1). We assume two levels, a spin up and a spin down, having the same energy ε, with the equilibrium chemical potential μ located right at ε, so that each level is half-filled since the Fermi function $f_0(E = \mu)$ equals 0.5. Based on what we have discussed so far we would expect a high conductance since the electrochemical potential lies right in the middle of each broadened level as shown in the upper sketch in Fig. 22.1.

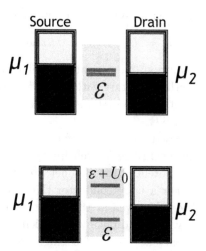

Fig. 22.1 The "bottom-up" view of Coulomb blockade: A two-level channel can show significantly lower density of states around $E = \mu$, and hence a higher resistance, if U_0 is large.

However, if the single electron charging energy U_0 is large then the picture could change to the lower one where one level floats up by U_0 due to the electron occupying the other level. Why doesn't the other level float up as well? Because no level feels any potential due to itself. This self-interaction correction is missed in the self-consistent field (SCF) model discussed in Chapter 8 where we wrote $U = U_0 N$. Instead we need an unrestricted SCF where each level i is not restricted to feeling the same potential. Instead it feels a potential U_i that depends on the change in the number of electrons occupying all levels except for i:

$$U_i = U_0 \left(N - N_i \right). \tag{22.1}$$

If we were to use Eq. (22.1) instead of $U = U_0 N$ we would obtain a picture like the lower one in Fig. 22.1, assuming that μ is adjusted to have approximately one electron inside the channel. We would find a self-consistent solution with

$$N_{dn} = 1, \ U_{up} = U_0, \ N_{up} = 0, \text{ and } U_{dn} = 0.$$

The down level will be occupied ($N_{dn} = 1$) and the resulting potential ($U_{up} = U_0$) will cause the up level to float up and be unoccupied ($N_{up} = 0$). Because it is unoccupied, the potential felt by the down level is zero ($U_{dn} = 0$), so that it does not float up, consistent with what we assumed to start with.

Of course, the solution with up and down interchanged

$$N_{up} = 1, \ U_{dn} = U_0, \ N_{dn} = 0, \ \text{and} \ U_{up} = 0$$

is also an equally valid solution. Numerically we will converge to one or the other depending on whether we start with an initial guess that has more N_{up} or N_{dn}. Experimentally the system will fluctuate between the two solutions randomly over time.

Why have we not worried about this before? Because it is not observable unless the charging energy U_0 is well in excess of both kT and the broadening. U_0/q is the potential the channel would float to if one electron were added to it. For a large conductor this potential is microvolts or smaller and is unobservable even at the lowest of temperatures. After all, any feature in energy is spread out by kT which is ~ 25 meV at room temperature and ~ 200 μeV at ~ 1 K. The single electron charging effect that we are talking about, becomes observable at least at low temperatures, once the conductor is small enough to make U_0 of the order of a meV. For molecular sized conductors, U_0 can be hundreds of meV making these effects observable even at room temperature.

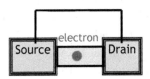

Fig. 22.2   The single electron charging energy U_0 is the electrostatic energy associated with one extra electron in the channel.

However, there is a second factor that also limits the observability of this effect. We saw in Chapter 18 that in addition to the temperature broadening $\sim kT$, there is a second and more fundamental broadening, $\gamma \sim h/t$ related to the transfer time. Single electron charging effects will be observed only if the Coulomb gap U_0 exceeds this broadening: $U_0 \gg h/t$. For this reason we would not expect to see this effect even in the smallest conductors, as long as it has good contacts.

22.1.1   *Current versus voltage*

Let us now move on from the low bias conductance to the full current-voltage characteristics of the two-level resistor. For simplicity we will assume that the levels remain fixed with respect to the source and are unaffected by the drain voltage, so that we do not have to worry about the kind of issues related to simple electrostatics that we discussed in Chapter 8.

A simple treatment ignoring electron-electron interactions then gives the curve marked *"non-interacting"* in Fig. 22.3. Once the electrochemical potential μ_2 crosses the levels at ε, the current steps up to its maximum value.

If we now include charging effects through a self-consistent potential $U = U_0 N$, the current step stretches out over a voltage range of $\sim U_0/q$, since the charging of the levels makes them float up and it takes more voltage to cross them completely.

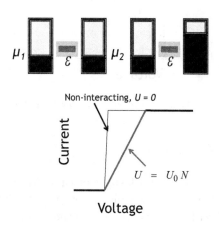

Fig. 22.3   Current-voltage characteristic of a two-level resistor with $U = 0$ and with $U = U_0 N$.

But if we include an SCF with self-interaction correction (Eq. (22.1)) we calculate a current-voltage characteristic with an intermediate plateau as shown in Fig. 22.4 which can be understood in terms of the energy level diagrams shown. At first only the lower level conducts giving only half the maximum current and only when the voltage is large enough for μ_2 to cross $\varepsilon + U_0$ that we get the full current.

Such intermediate plateaus in the *I-V* characteristics have indeed been observed but the details are not quite right. The correct plateau current is believed to be $2/3$ and not $1/2$ of the total current of $2q/t$. This represents an effect that is difficult to capture within a one-electron picture, though it can be understood clearly if we adopt a different approach altogether, which we will now describe.

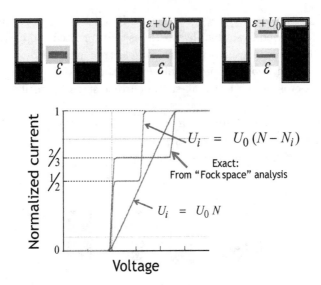

Fig. 22.4 Current-voltage characteristic of a two-level resistor: Exact and with two different SCF potentials.

22.2 Fock Space Description

This approach is based on the Fock space picture introduced in Chapter 15. As we discussed earlier, in this new picture we do not think in terms of one-electron levels that get filled or emptied from the contacts. Instead we think in terms of the system being driven from one state to another.

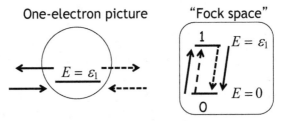

Fig. 22.5 One-electron picture versus Fock space picture for a one-level channel.

For example Fig. 22.5 shows how we would view the one-level resistor in this Fock space picture where the system can be one of two states: 0 representing an empty state, and 1 representing a full state. Figure 22.6 shows the two pictures for a two-level resistor. In general an N-level resistor will have 2^N Fock space states.

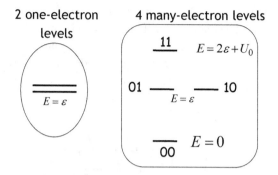

Fig. 22.6   One-electron picture versus Fock space picture for a two-level channel.

22.2.1   *Equilibrium in Fock space*

As we discussed in Chapter 15, there is a well-defined procedure for finding the probabilities of finding the system in a given eigenstate i at equilibrium.

$$p_i = \frac{1}{Z} e^{-(E_i - \mu N_i)/kT} \quad \text{(Same as Eq. (15.18)).}$$

We could use this to calculate any equilibrium property. For example suppose we want to find the number of electrons, n occupying the two-level channel shown in Fig. 22.6 if it is in equilibrium with an electrochemical potential μ.

Figure 22.7 shows the result obtained by plotting n versus μ from the equation

$$n = \sum_i N_i p_i = p_{01} + p_{10} + 2p_{11}$$

using the equilibrium probabilities from Eq. (15.18) cited above. Note how the electron number changes by one as μ crosses ε and then again when μ crosses $\varepsilon + U_0$ in keeping with the lower picture in Fig. 22.1.

Note, however, that we did not assume the picture from Fig. 22.1 with two one-electron states at different energies. We assumed two one-electron states with the same energy (Fig. 22.6) but having an interaction energy that is included in the Fock space picture.

If we are interested in the *low bias conductance*, G as a function of μ, we could deduce it from the $n(\mu)$ plot in Fig. 22.7. As we discussed in Chapter 2, current flow is essentially because the two contacts with different μ's have different agendas, since one likes to see more electrons in the channel than the other. From this point of view one could argue that the conductance should be proportional to $dn/d\mu$ and show peaks at $\mu = \varepsilon$ and at $\mu = \varepsilon + U_0$ as shown. This is indeed what has been observed experimentally for the low bias conductance of small conductors in the single-electron charging regime where U_0 exceeds both the thermal energy kT and the energy broadening due to contacts.

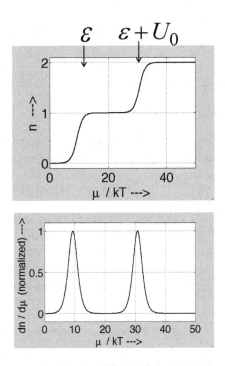

Fig. 22.7   Equilibrium number of electrons, n in the two-level channel shown in Fig. 22.6 as a function of μ, assuming $\varepsilon = 10\,kT$, $U_0 = 20\,kT$. The conductance can be argued to be proportional to the derivative $dn/d\mu$ showing peaks when μ equals ε and $\varepsilon + U_0$. For MATLAB script, see Appendix H.3.1.

As we saw in Chapter 5, low bias conductance is an equilibrium property that can be deduced using the principles of equilibrium statistical mechanics. Current flow at higher voltages on the other hand requires the methods of non-equilibrium statistical mechanics. Let me explain briefly how one could understand the 2/3 plateau shown in Fig. 22.4 by calculating the current at high bias in the Fock space picture.

22.2.2 *Current in the Fock space picture*

To calculate the current we write an equation for the probability that the system will be found in one of its available states, which must all add up to one. For example for the one level resistor we could write

$$\nu_1 p_0 = \nu_2 p_1 \quad \rightarrow \quad \frac{p_1}{p_0} = \frac{\nu_1}{\nu_2} \quad \rightarrow \quad p_1 = \frac{\nu_1}{\nu_1 + \nu_2}$$

assuming that the left contact sends the system from the 0 state to the 1 state at a rate ν_1, while the right contact takes it in the reverse direction at a rate ν_2 and at steady-state the two must balance. The current is given by

$$I = q\nu_2 p_1 = q \frac{\nu_1 \nu_2}{\nu_1 + \nu_2} \tag{22.2}$$

in agreement with our earlier result in Chapter 18 (see Eq. (18.10b)) obtained from a one-electron picture.

But the real power of this approach is evident when we consider levels with multiple interacting levels. Consider for example the two-level resistor biased such that electrons can come in from the left contact and transfer the system from 00 to 01 or to 10, but not to the 11 state because of the high charging energy U_0. This is the biasing condition that leads to a plateau at 2/3 the maximum value (Fig. 22.4) that we mentioned earlier.

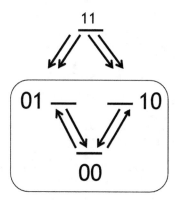

In this biasing condition, the system can only come out of the 11 state, but never transfer into it, and so the steady-state condition can be calculated simply by considering the kinetics of the three remaining states in

Fock space, namely 00, 01 and 10:

$$2\nu_1 p_{00} = \nu_2(p_{01} + p_{10})$$

$$\rightarrow \quad \frac{p_{01} + p_{10}}{p_{00}} = \frac{2\nu_1}{\nu_2}$$

$$\rightarrow \quad p_{01} + p_{10} = \frac{2\nu_1}{2\nu_1 + \nu_2}$$

where we have made use of the requirement that all three probabilities must add up to one. Hence

$$I = q\nu_2(p_{01} + p_{10}) = q\frac{2\nu_1\nu_2}{2\nu_1 + \nu_2}$$

with

$$\nu_1 = \nu_2 \quad \rightarrow \quad I = \frac{2}{3}q\nu_1$$

which is 2/3 the maximum current as stated earlier.

It is important to note the very special nature of the solution we just obtained which makes it hard to picture within a one-electron picture. We showed that the system is equally likely to be in the states 00, 01 and the 10 states, but zero probability of being in the 11 state.

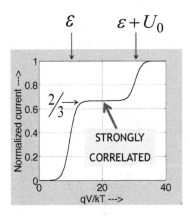

Fig. 22.8   The intermediate plateau in the current corresponds to the channel being in a strongly correlated state. For MATLAB script, see Appendix H.3.2.

In other words, if we looked at the up-spin or the down-spin state (in the one-electron picture) we would find them occupied with 1/3 probability. If electrons were independent then we would expect the probability for both to be occupied to be the product = 1/9.

Instead it is zero, showing that the electrons are correlated and cannot be described with a one-electron occupation factor f of the type we have been using throughout this book. Even with quantum transport we replaced the f's with a matrix $\mathbf{G}^n$ obtained by summing the $\boldsymbol{\psi}\boldsymbol{\psi}^\dagger$ for individual electrons. This adds sophistication to our understanding of the one-electron state, but it still does not tell us anything about two-electron correlations.

22.3   Entangled States

What we just saw with one quantum dot is actually just the proverbial tip of the iceberg. Things get more interesting if we consider two or more quantum dots.

For example, with two coupled quantum dots we could write the one-electron Hamiltonian matrix as a (4×4) matrix using the up and down states in dots 1 and 2 as the basis functions as follows:

$$
\mathbf{H} \;=\;
\begin{array}{c}
\\
u_1 \\ u_2 \\ d_1 \\ d_2
\end{array}
\begin{array}{c}
\begin{array}{cccc} u_1 & u_2 & d_1 & d_2 \end{array} \\
\left[
\begin{array}{cccc}
\varepsilon_1 & t & 0 & 0 \\
t & \varepsilon_2 & 0 & 0 \\
0 & 0 & \varepsilon_1 & t \\
0 & 0 & t & \varepsilon_2
\end{array}
\right]
\end{array}
\tag{22.3}
$$

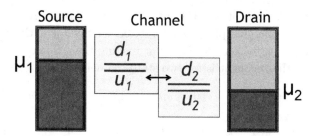

But what are the Fock space states? With four one-electron states we expect a total of $2^4 = 16$ Fock space states, containing 0, 1, 2, 3 or 4 electrons. The number of n-electron states in Fock space is given by ^4C_n: one with $n = 0$, four with $n = 1$, six with $n = 2$, four with $n = 3$ and one with $n = 4$.

If there were no inter-dot coupling then these sixteen states would be the eigenstates and we could analyze their dynamics in Fock space just as we did for one dot. But in the presence of inter-dot coupling the true eigenstates

4-electron $\boxed{1111}$

3-electron $\boxed{1110, 1101, 1011, 0111}$

2-electron $\boxed{1100, 1010, 1001, 0110, 0101, 0011}$

1-electron $\boxed{1000, 0100, 0010, 0001}$

0-electron $\boxed{0000}$

are linear combinations of these states and these entangled states can lead to novel physics and make it much more interesting.

The 0-electron and 4-electron states are trivially composed of just one Fock space state, while the 1-electron state is essentially the same as the states in a one-electron picture. Indeed the 3-electron state also has a structure similar to the one-electron state and could be viewed as a 1-hole state.

The 2-electron states, however, have an interesting non-trivial structure. Consider the six 2-electron states which we label in terms of the two states that are occupied: $u_1 d_1$, $u_2 d_2$, $u_1 d_2$, $u_2 d_1$, $u_1 u_2$, $d_1 d_2$. Using these we can write the Fock space Hamiltonian **HH** as explained below.

The *diagonal* elements of **HH** are written straightforwardly by adding the one-electron energies plus an interaction energy U_0 if the two basis functions happen to be on the same dot making their Coulomb repulsion much stronger than what it is for two states on neighboring dots.

$$\mathbf{HH} = \begin{array}{c} \\ u_1 d_1 \\ u_2 d_2 \\ u_1 d_2 \\ u_2 d_1 \\ u_1 u_2 \\ d_1 d_2 \end{array} \begin{array}{cccccc} u_1 d_1 & u_2 d_2 & u_1 d_2 & u_2 d_1 & u_1 u_2 & d_1 d_2 \\ \left[\begin{array}{cccccc} 2\varepsilon_1 + U_0 & 0 & t & t & 0 & 0 \\ 0 & 2\varepsilon_2 + U_0 & t & t & 0 & 0 \\ t & t & \varepsilon_1 + \varepsilon_2 & 0 & 0 & 0 \\ t & t & 0 & \varepsilon_1 + \varepsilon_2 & 0 & 0 \\ 0 & 0 & 0 & 0 & \varepsilon_1 + \varepsilon_2 & 0 \\ 0 & 0 & 0 & 0 & 0 & \varepsilon_1 + \varepsilon_2 \end{array} \right] \end{array} \quad (22.4)$$

The *off-diagonal* entries t are obtained by noting that this quantity couples the one electron states u_1 to u_2 and d_1 to d_2. With two electron states we have inserted t for non-diagonal elements that couples those states for which one state remains unchanged while the other changes from u_1 to u_2 or from d_1 to d_2.

The lowest eigenstate obtained from the two-electron Hamiltonian in Eq. (22.4) is with a wavefunction of the form (s_1, $s_2 < 1$)

$$S: (\{u_1d_2\} + \{u_2d_1\}) + s_1\{u_1d_1\} + s_2\{u_2d_2\} \qquad (22.5)$$

is called the *singlet state*. Next comes a set of three states (called the *triplets*) that are higher in energy. These have the form

$$T_1: \frac{1}{\sqrt{2}}(\{u_1d_2\} - \{u_2d_1\})$$
$$T_3: \{u_1u_2\}$$
$$T_3: \{d_1d_2\}. \qquad (22.6)$$

A system with two electrons is normally viewed as occupying two one-electron states. The states T_2 and T_3 permit such a simple visualization. But the states S and T_1 do not.

For example, each term in the state

$$T_1: \frac{1}{\sqrt{2}}(\{u_1d_2\} - \{u_2d_1\})$$

permits a simple visualization: $\{u_1d_2\}$ stands for an upspin electron in 1 and a downspin electron in 2 while $\{u_2d_1\}$ represents an upspin in 2 and a downspin in 1. But the real state is a superposition of these two "simple" or *unentangled* states and there is no way to define two one-electron states a and b such that the two-electron state could be viewed as $\{ab\}$. Such states are called entangled states which comprise the key entity in the emerging new field of quantum information and computing.

How would we compute the properties of such systems? The equilibrium properties are still described by the general law of equilibrium stated earlier

$$p_i = \frac{1}{Z}e^{-(E_i-\mu N_i)/kT} \quad \text{(Same as Eq. (15.18))}$$

and using the equilibrium properties to evaluate the average number of electrons

$$n = \sum_i N_i p_i.$$

The energies E_i are obtained by diagonalizing the Fock space Hamiltonian **HH** that we just discussed. Figure 22.9 shows the plot of n versus μ which looks like Fig. 22.7, but the middle plateau now involves the entangled singlet state just discussed. There is also some additional structure that we will not get into. The main point we wanted to make is that the law of equilibrium statistical mechanics is quite general and can be used in this case.

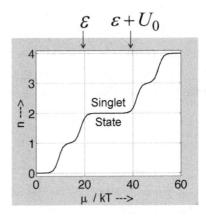

Fig. 22.9   Equilibrium number of electrons, n in the two-level channel shown in Fig. 22.6 as a function of μ, assuming $\varepsilon = 10\,kT$, $U_0 = 20\,kT$. For MATLAB script, see Appendix H.3.3.

But the calculation of current at high bias is a non-equilibrium problem that is not as straightforward. Using the entangled states one could set up a rate equation as we did in the last section and understand some of the interesting effects that have been observed experimentally including negative differential resistance (NDR), that is a decrease in current with increasing voltage (see for example Muralidharan *et al.*, 2007). More generally one needs quantum rate equations to go beyond the simple rate equations we discussed and handle coherences (Braun *et al.*, 2004, Braig and Brouwer, 2005).

Can we model transport involving correlated and/or entangled states exactly if we use a Fock space picture instead of using NEGF and including interactions only approximately through self-energies? Sort of, but not quite.

There are two problems. The first is practical. A N-level problem in the one-electron picture escalates into a 2^N level problem n the Fock space picture. The second is conceptual.

We saw in Chapter 18 how the NEGF method allows us to include quantum broadening in the one-electron Schrödinger equation. To our knowledge there is no comparable accepted method for including broadening in the Fock space picture. So the rate equation approach from the last section works fine for weakly coupled contacts where the resulting broadening is negligible, but the regime with broadening comparable to the charging energy stands out as a major challenge in transport theory. Even the system with two levels (Fig. 22.7) shows interesting structure in $n(\mu)$ in this regime ("Kondo peak") that has occupied condensed matter physicists for many decades.

One could view Coulomb blockade as the bottom-up version of the Mott transition, a well-studied phenomenon in condensed matter physics. In a long chain of atoms, the levels ε and $\varepsilon + U_0$ (Fig. 22.1) will each broaden into a band of width $\sim 2t_0$, $t_0/\hbar$ being the rate at which electrons move from one atomic site to the next. These are known as the lower and upper Hubbard bands. If their separation U_0 exceeds the width $2t_0$ of each band we will have a Mott insulator where the electrochemical potential lies in the middle of the two bands with very low density of states and hence very low conductance. But if U_0 is small, then the two bands form a single half-filled band with a high density of states at $E = \mu_0$ and hence a high conductance.

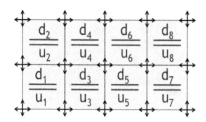

Needless to say, the full theory of the Hubbard bands is far more complicated than this oversimplified description might imply and it is one of the topics that has occupied condensed matter theorists for over half a century. Since the late 1980s it has acquired an added significance with the discovery of a new class of superconductors operating at relatively high temperatures above 100 K, whose mechanism continues to be controversial and hotly debated.

This problem remains one of the outstanding problems of condensed matter theory, but there seems to be general agreement that the essential physics involves a two-dimensional array of quantum dots with an inter-dot coupling that is comparable to the single dot charging energy.

PART 3
Spin Transport

Chapter 23

Rotating an Electron

Related video lecture available at course website, Unit 4: L4.1.

Back in Section 12.2 of Part A, we talked about the concept of spin potentials and how they can be generated in two ways, namely (Fig. 23.1)

(1) use of *magnetic contacts* with *ordinary channels* (Section 12.2.1),
(2) use of *ordinary contacts* with *spin-momentum locked channels* (Section 12.2.2).

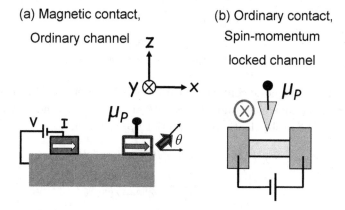

Fig. 23.1 Spin potentials can be generated (a) using magnetic contacts with ordinary channels, or (b) using ordinary contacts with spin-momentum locked channels. In either case a magnetic probe can be used to measure the spin potential in the channel.

In the first case the channel-contact interface presents different resistances to the two spins so that the surface potential is different for the two spins right under the contact, giving rise to a spin potential μ_S (see Eq. (12.18b),

Part A) in addition to the usual charge potential μ:

$$\mu = \frac{\mu^{up} + \mu^{dn}}{2} \tag{23.1a}$$

$$\mu_S = \frac{\mu^{up} - \mu^{dn}}{2}. \tag{23.1b}$$

This potential persists for a distance up to a spin coherence length from the contact, and can be measured by placing a magnetic probe within this range. In the second case the contacts are normal and the current flow in the channel gives rise to a separation in the potentials μ^+ and μ^- for right and left moving electrons. For a *2D spin-momentum locked channel* in the x-y plane, right $(+\hat{\mathbf{x}})$ and left $(-\hat{\mathbf{x}})$ moving electrons correspond to spins with positive and negative components along $\hat{\mathbf{z}} \times \hat{\mathbf{x}} = \hat{\mathbf{y}}$. A difference $(\mu^+ - \mu^-)$ thus translates into a spin potential $\mu_S \sim \mu^{up} - \mu^{dn}$, where *up* and *dn* refer to spins pointing along positive and negative y.

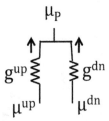

Fig. 23.2   Simple circuit model for a voltage probe.

How does a magnetic contact measure the spin potential? Through the difference in the interface conductances g^{up} and g^{dn} for up and down spins (Fig. 23.2). The probe floats to a potential μ_P such that

$$g_{up}(\mu_{up} - \mu_P) + g_{dn}(\mu_{dn} - \mu_P) = 0 \tag{23.2a}$$

$$\rightarrow \mu_P = \frac{g^{up}\mu^{up} + g^{dn}\mu^{dn}}{g^{up} + g^{dn}} \tag{23.2b}$$

which can be written in terms of μ, μ_S and the polarization P of the magnet:

$$\mu_P = \mu + P\mu_S \tag{23.3a}$$

$$P \equiv \frac{g^{up} - g^{dn}}{g^{up} + g^{dn}}. \tag{23.3b}$$

Note that if we reverse a magnet, $g^{up} \leftrightarrow g^{dn}$, so that $P \to -P$. This means that if we reverse a magnet from $+\hat{\mathbf{m}}$ to $-\hat{\mathbf{m}}$, the probe potential will change by

$$\mu_P(+\hat{\mathbf{m}}) - \mu_P(-\hat{\mathbf{m}}) = 2P\mu_S. \tag{23.4}$$

Indeed this is the standard method for measuring spin potentials, namely by looking at the change in the probe potential when the magnet is reversed.

What voltage would the magnetic probe measure if it were neither parallel nor anti-parallel to the spins, but instead made some arbitrary angle with it (Fig. 23.1a)? The answer can be stated quite simply:

$$\mu_P = \mu + \mathbf{P} \cdot \boldsymbol{\mu}_S \tag{23.5}$$

where $\mathbf{P} = P\hat{\mathbf{m}}$ and the spin potential is a vector pointing in the direction that defines "up". The derivation of Eq. (23.5) should be clearer as we discuss spins in more depth in this chapter.

23.1 Polarizers and Analyzers

How do we understand the general result in Eq. (23.5)? For those unfamiliar with electron spin, the simplest analogy is probably that of photon polarization. As we learn in freshman physics, a polarizer-analyzer combination lets through a flux proportional to $\cos^2 \theta$. It is maximum when the two are parallel ($\theta = 0°$), and a minimum when the two are perpendicular ($\theta = 90°$).

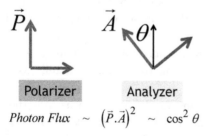

$$\text{Photon Flux} \quad \sim \quad \left(\vec{P}.\vec{A}\right)^2 \quad \sim \quad \cos^2 \theta$$

What about electrons? Suppose we have electrons that are all up, so that $\mu_{dn} = 0$, and from Eq. (23.1), we have $|\mu_S| = \mu = \mu^{up}/2$. Equation (23.5) then gives

$$\frac{\mu_P}{\mu^{up}} = \frac{1 + P\cos\theta}{2}.$$

As with photons, the voltage is a maximum when the probe (analyzer) is parallel to the electron polarization ($\theta = 0°$). But with electrons the

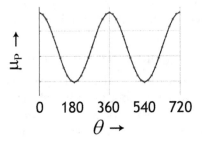

Fig. 23.3   Probe voltage variation as its magnetization is rotated.

minimum occurs, not when the two are perpendicular ($\theta = 90°$) but when the two are antiparallel ($\theta = 180°$) as shown in Fig. 23.3.

Indeed if we assume a perfect voltage probe with $P = 1$, we have

$$\frac{\mu_P}{\mu} = \frac{1 + \cos\theta}{2} = \cos^2\frac{\theta}{2}$$

showing that the analyzer lets through a fraction of electrons proportional to

$$\underbrace{\cos^2\left(\frac{\theta}{2}\right)}_{\text{Electrons}} \quad \text{rather than} \quad \underbrace{\cos^2\theta}_{\text{Photons}}.$$

One point that causes some confusion is the following. It seems that if we had electrons in the channel whose spin direction we did not know, we could measure it using a magnet. As we turn the magnet the measured voltage should go through maxima and minima as shown in Fig. 23.3, and the direction corresponding to a maximum tells us the direction of the electron spin.

But doesn't quantum mechanics teach us that the spin of an electron cannot be exactly measured? Yes, but that is true if we had just one electron. Here we are talking of an "army" of electrons identically prepared by an injecting contact and what our magnet measures is the average over many many such electrons. This is not in violation of any basic principle.

Anyway, the bottom line is that for electron spin, orthogonal directions are not represented by say z and x that are 90 degrees apart. Rather they are represented by '*up*' and '*down*' that are 180 degrees apart. And that is why a proper description of electron spin requires spinors rather than vectors as we will now discuss.

A vector $\hat{n}$ is described by three real components, namely the components along x, y and z, but spinors are described by two complex components, which are its components along up and down:

$$\underbrace{\begin{Bmatrix} n_x \\ n_y \\ n_z \end{Bmatrix}}_{\text{Vector}}, \quad \underbrace{\begin{Bmatrix} \psi_{up} \\ \psi_{dn} \end{Bmatrix}}_{\text{Spinor}}.$$

Nevertheless we visualize the spinor as an object pointing in some direction just like a vector. How do we reconcile the visual picture with the 2-component complex representation?

A spinor pointing along a direction described by a unit vector

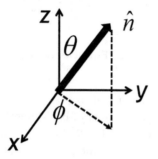

$$\hat{n} \equiv \begin{Bmatrix} \sin\theta \, \cos\phi \\ \sin\theta \, \sin\phi \\ \cos\theta \end{Bmatrix} \tag{23.6}$$

has components given by

$$\begin{Bmatrix} \cos\left(\dfrac{\theta}{2}\right) e^{-i\phi/2} \equiv c \\[2ex] \sin\left(\dfrac{\theta}{2}\right) e^{+i\phi/2} \equiv s \end{Bmatrix}. \tag{23.7}$$

This is of course not obvious and later in the chapter I will try to explain why Eqs. (23.6) and (23.7) represent isomorphic (more correctly "homomorphic") ways to represent an abstract rotatable object pointing in some direction. For the moment let us accept Eq. (23.7) for the components of a spinor and work out some of its consequences.

23.2   Spin in NEGF

Although these subtleties of visualization and interpretation take some time to get used to, formally it is quite straightforward to incorporate spin into the quantum transport formalism from Chapter 18. The basic equations from Eq. (18.1) through (18.4) remain the same, but all the matrices like **H**, **Σ**, **G**^{n}, and **A** *become twice as big* (Fig. 23.4).

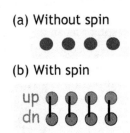

Fig. 23.4   Inclusion of spin in NEGF doubles the number of "grid points" or basis functions.

Ordinarily these matrices are of size $(N \times N)$, if N is the number of grid points (or more formally the number of basis functions) used to describe the channel. Inclusion of spin basically doubles the number of basis functions: every grid point turns into two points, an up and a down (Fig. 23.4).

How would we write down **H** including spin? We can visualize the TB parameters (See Fig. 17.7) exactly as before except that each on-site element α and the coupling elements β are each (2×2) matrices (Fig. 23.5). In the

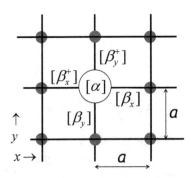

Fig. 23.5   2D Lattice with each element given by a (2×2) matrix to reflect spin-related properties.

simplest case, we can imagine a "spin-innocent" channel that treats both spin components identically. Such a channel can be modeled by choosing the TB parameters as follows:

$$\boldsymbol{\alpha} = \varepsilon\mathbf{I}, \quad \boldsymbol{\beta}_x = t\mathbf{I}, \quad \boldsymbol{\beta}_y = t\mathbf{I} \tag{23.8}$$

where $\mathbf{I}$ is the (2×2) identity matrix. We effectively have two identical decoupled Hamiltonians that includes no new physics.

Similarly we can write the self-energy $\boldsymbol{\Sigma}$ for ordinary contacts that treat both spin components identically simply by taking our usual values and multiplying by $\mathbf{I}$. This would again be in the category of a trivial extension that introduces no new physics. The results should be the same as what we would get if we worked with one spin only and multiplied by two at the end.

All spin-related phenomena like the ones we discussed in Chapter 12 arise either from non-trivial contacts described by $\boldsymbol{\Sigma}$ with spin-related properties or from channels described by $\mathbf{H}$ with spin-related properties or both.

Let us now try to get a feeling for spin transport problems by applying the NEGF method to a series of examples, starting with a simple one-level version of the spin valve we started Chapter 12 with. From a *computational* point of view the only question is how to write down $\mathbf{H}$ and $\boldsymbol{\Sigma}$. Once we have these, the rest is standard. One can then proceed to *understand and enjoy* the physics.

23.3 One-level Spin Valve

Related video lecture available at course website, Unit 4: L4.2.

As we discussed in Chapter 12, a spin valve (Fig. 23.6) shows different conductances G_P and G_{AP} depending on whether the magnetic contacts have parallel (P) or anti-parallel (AP) magnetizations. Using a simple model we showed that the *magnetoresistance* (MR) can be expressed as

$$MR \equiv \frac{G_P}{G_{AP}} - 1 = \frac{P^2}{1 - P^2}$$

where the polarization P was defined in terms of the interface resistances. In that context we noted that the standard expression for the MR for magnetic tunnel junctions (MTJ's) has an extra factor of two

$$MR \equiv \frac{G_P}{G_{AP}} - 1 = \frac{2P^2}{1 - P^2} \tag{23.9}$$

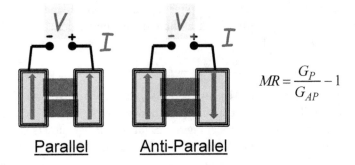

$$MR = \frac{G_P}{G_{AP}} - 1$$

Fig. 23.6   The spin-valve has different conductances G_P and G_{AP} depending on whether the magnetic contacts have parallel (P) or anti-parallel (AP) magnetization.

which could be understood if we postulated that the overall resistance was proportional to the product of the interface resistances and not their sum.

We could obtain this result including the factor of two directly from our NEGF model if we apply it to a one-level resistor and assume that the equilibrium electrochemical potential μ_0 is located many kT's below the energy ε of the level as sketched.

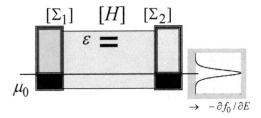

Figure 23.7 summarizes the (2×2) matrices $\mathbf{H}$ and $\mathbf{\Sigma}$ for this device. Also shown for comparison are the corresponding (1×1) "matrices" (that is, just numbers) for the same device without spin. Note that the channel is assumed to treat both spins identically so that $\mathbf{H}$ is essentially an identity matrix, but the $\mathbf{\Sigma}$'s have different values for the up and downspin components.

Using these matrices it is straightforward to obtain the Green's function

$$G^R = \begin{bmatrix} E - \varepsilon + \frac{i}{2}(\gamma_{1u} + \gamma_{2u}) & 0 \\ 0 & E - \varepsilon + \frac{i}{2}(\gamma_{1d} + \gamma_{2d}) \end{bmatrix}^{-1}$$

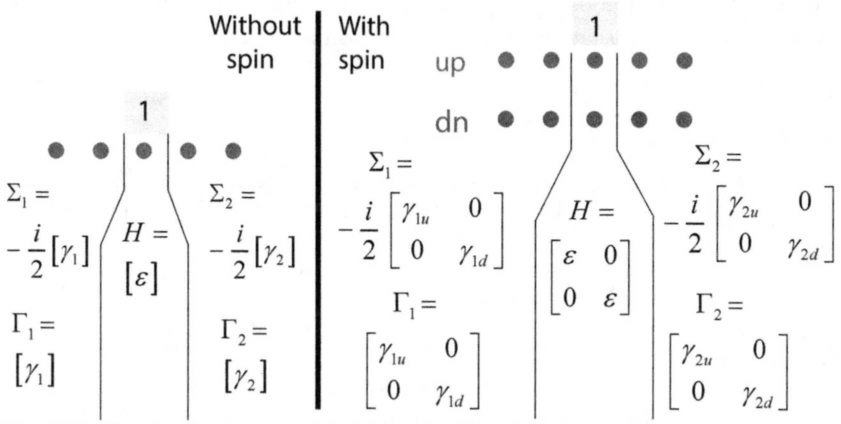

Fig. 23.7 One-level spin-valve: Modifying the **H** and **Σ** for a spin-less one-level device to represent a one-level spin valve.

and hence the transmission

$$\bar{T} = \text{Trace}\left[\mathbf{\Gamma}_1 \mathbf{G}^R \mathbf{\Gamma}_2 \mathbf{G}^A\right]$$

$$= \frac{\gamma_{1u}\gamma_{2u}}{(E-\varepsilon)^2 + \left(\dfrac{\gamma_{1u}+\gamma_{2u}}{2}\right)^2} + \frac{\gamma_{1d}\gamma_{2d}}{(E-\varepsilon)^2 + \left(\dfrac{\gamma_{1d}+\gamma_{2d}}{2}\right)^2}. \qquad (23.10)$$

For the **parallel** (P) configuration we can assume both contacts to be identical so that we can write ($\alpha > \beta$)

$$\gamma_{1u} = \gamma_{2u} \equiv \alpha \qquad (23.11\text{a})$$

$$\gamma_{1d} = \gamma_{2d} \equiv \beta \qquad (23.11\text{b})$$

while for the **anti-parallel** (AP) configuration the second contact has the roles of α and β reversed with respect to the former:

$$\gamma_{1u} = \gamma_{2d} \equiv \alpha \qquad (23.12\text{a})$$

$$\gamma_{1d} = \gamma_{2u} \equiv \beta. \qquad (23.12\text{b})$$

Inserting Eqs. (23.11) into Eq. (23.10) we have the P - transmission

$$\bar{T}_P = \frac{\alpha^2}{(E-\varepsilon)^2 + \alpha^2} + \frac{\beta^2}{(E-\varepsilon)^2 + \beta^2}$$

while using Eqs. (23.12) in Eq. (23.10) gives the AP - transmission

$$\bar{T}_{AP} = \frac{2\alpha\beta}{(E-\varepsilon)^2 + \left(\dfrac{\alpha+\beta}{2}\right)^2}.$$

The measured conductance depends on the average transmission over a range of energies of a few kT around μ_0. Assuming that

$$\varepsilon - \mu_0 \gg kT, \alpha, \beta$$

we can write

$$G_P \sim \bar{T}_P(E = \mu_0) = \frac{\alpha^2}{(\mu_0 - \varepsilon)^2 + \alpha^2} + \frac{\beta^2}{(\mu_0 - \varepsilon)^2 + \beta^2} \approx \frac{\alpha^2 + \beta^2}{(\mu_0 - \varepsilon)^2}$$

and

$$G_{AP} \sim \bar{T}_{AP}(E = \mu_0) \approx \frac{2\alpha\beta}{(\mu_0 - \varepsilon)^2}.$$

This gives us

$$MR \equiv \frac{G_P}{G_{Ap}} - 1 = \frac{\alpha^2 + \beta^2}{2\alpha\beta} - 1 = \frac{2P^2}{1 - P^2}$$

as stated earlier in Eq. (23.9) with the polarization defined as

$$P \equiv \frac{\alpha - \beta}{\alpha + \beta}. \tag{23.13}$$

Actually we could also obtain the result without the factor of 2, obtained from the resistor model in Chapter 12, if we assume that μ_0 is located right around the level ε, with $kT \gg \alpha$ and $kT \gg \beta$. But we leave that as an exercise. After all this is just a toy problem intended to get us started.

23.4 Rotating Magnetic Contacts

Related video lecture available at course website, Unit 4: L4.3.

We argued in the last section that for an anti-parallel spin valve, the second contact simply has the roles of α and β reversed relative to the first, so that we can write

$$\Gamma_1 = \begin{bmatrix} \alpha & 0 \\ 0 & \beta \end{bmatrix}, \quad \Gamma_2 = \begin{bmatrix} \beta & 0 \\ 0 & \alpha \end{bmatrix}.$$

But how would we write the corresponding matrix for a contact if it were pointing along some arbitrary direction defined by a unit vector $\hat{n}$? The answer is

$$\Gamma = \frac{\alpha + \beta}{2} I + \frac{\alpha - \beta}{2} \begin{bmatrix} n_z & n_x - i\,n_y \\ n_x + i\,n_y & -n_z \end{bmatrix} \tag{23.14}$$

where n_x, n_y and n_z are the components of the unit vector $\hat{n}$ along x, y and z respectively. This result is of course not obvious and we will try to justify it shortly. But it is reassuring to note that the results for both the parallel and the anti-parallel contact come out as special cases of this general result (Eq. (23.14)):

$$\text{If} \quad n_z = +1, \ n_x = n_y = 0: \quad \boldsymbol{\Gamma} = \begin{bmatrix} \alpha & 0 \\ 0 & \beta \end{bmatrix}$$

$$\text{If} \quad n_z = -1, \ n_x = n_y = 0: \quad \boldsymbol{\Gamma} = \begin{bmatrix} \beta & 0 \\ 0 & \alpha \end{bmatrix}.$$

One way to understand where Eq. (23.14) comes from is to note that the appropriate matrix describing a magnet pointing along $\hat{n}$ would be

$$\tilde{\boldsymbol{\Gamma}} = \begin{bmatrix} \alpha & 0 \\ 0 & \beta \end{bmatrix} \tag{23.15}$$

if we were to take $+\hat{n}$ and $-\hat{n}$ as our reference directions instead of $+\hat{z}$ and $-\hat{z}$ as we normally do. How could we then transform the $\tilde{\boldsymbol{\Gamma}}$ from Eq. (23.15) into the usual $\pm\,\hat{z}$ basis?

Answer: Transform from the $\pm\,\hat{n}$ to the $\pm\,\hat{z}$ basis

$$
\begin{array}{cc}
\hat{n} \quad -\hat{n} \\
\begin{matrix}\hat{z} \\ -\hat{z}\end{matrix}
\underbrace{\begin{bmatrix} c & -s^* \\ s & c^* \end{bmatrix}}_{\mathbf{V}}
\end{array}
\quad
\begin{array}{cc}
\hat{n} \quad -\hat{n} \\
\begin{matrix}\hat{n} \\ -\hat{n}\end{matrix}
\begin{bmatrix} \alpha & 0 \\ 0 & \beta \end{bmatrix}
\end{array}
\quad
\begin{array}{cc}
\hat{z} \quad -\hat{z} \\
\begin{matrix}\hat{n} \\ -\hat{n}\end{matrix}
\underbrace{\begin{bmatrix} c^* & s^* \\ -s & c \end{bmatrix}}_{\mathbf{V}^\dagger}
\end{array}
\tag{23.16}
$$

using the unitary transformation matrix $\mathbf{V}$ whose columns represent the components of a spinor pointing along $\pm\hat{n}$. The first column follows from the result we stated earlier in Eq. (23.7), while the second can be obtained from Eq. (23.7) if we set

$$\theta \rightarrow \pi - \theta, \ \phi \rightarrow \pi + \phi$$

and remove a common phase factor from the two components.

Multiplying out the three matrices in Eq. (23.16) we have

$$
\boldsymbol{\Gamma} = \begin{bmatrix} c & -s^* \\ s & c^* \end{bmatrix} \begin{bmatrix} \alpha & 0 \\ 0 & \beta \end{bmatrix} \begin{bmatrix} c^* & s^* \\ -s & c \end{bmatrix} = \begin{bmatrix} c & -s^* \\ s & c^* \end{bmatrix} \begin{bmatrix} \alpha c^* & \alpha s^* \\ -\beta s & \beta c \end{bmatrix}
$$

$$
= \begin{bmatrix} \alpha cc^* + \beta ss^* & (\alpha - \beta)\, cs^* \\ (\alpha - \beta)\, sc^* & \alpha ss^* + \beta cc^* \end{bmatrix}.
$$

Making use of the definitions of c and s from Eq. (23.7) and some common trigonometric identities like

$$2\cos^2\left(\frac{\theta}{2}\right) = 1 + \cos\theta,$$

$$2\sin^2\left(\frac{\theta}{2}\right) = 1 - \cos\theta, \text{ and} \qquad (23.17)$$

$$2\sin\left(\frac{\theta}{2}\right)\cos\left(\frac{\theta}{2}\right) = \sin\theta$$

we can rewrite this as

$$\Gamma = \frac{1}{2}\begin{bmatrix} (\alpha+\beta)+(\alpha-\beta)\cos\theta & (\alpha-\beta)\sin\theta\,e^{-i\phi} \\ (\alpha-\beta)\sin\theta\,e^{+i\phi} & (\alpha+\beta)-(\alpha-\beta)\cos\theta \end{bmatrix}$$

which leads to the result stated earlier in Eq. (23.14) if we make use of Eq. (23.6) for the x, y and z components of a unit vector.

Finally let me note that if we define the polarization as a vector whose magnitude is given by Eq. (23.13) and direction is given by $\hat{n}$:

$$\mathbf{P} \equiv P\,\hat{n} = \frac{\alpha-\beta}{\alpha+\beta}\,\hat{n} \qquad (23.18)$$

then we could rewrite Eq. (23.14) as

$$\Gamma = \frac{\alpha+\beta}{2}\left(\mathbf{I} + \begin{bmatrix} P_z & P_x - i\,P_y \\ P_x + i\,P_y & -P_z \end{bmatrix}\right) \qquad (23.19)$$

which can be rearranged as shown

$$\frac{\Gamma}{(\alpha+\beta)/2} = \underbrace{\begin{bmatrix} 1 & 0 \\ 0 & 1 \end{bmatrix}}_{\mathbf{I}} + P_x\underbrace{\begin{bmatrix} 0 & 1 \\ 1 & 0 \end{bmatrix}}_{\sigma_x} + P_y\underbrace{\begin{bmatrix} 0 & -i \\ +i & 0 \end{bmatrix}}_{\sigma_y} + P_z\underbrace{\begin{bmatrix} 1 & 0 \\ 0 & -1 \end{bmatrix}}_{\sigma_z}.$$

Any (2×2) matrix can be expressed in terms of the four matrices appearing here consist of the identity matrix $\mathbf{I}$ along with the three **Pauli spin matrices**

$$\sigma_x \equiv \begin{bmatrix} 0 & 1 \\ 1 & 0 \end{bmatrix}, \quad \sigma_y \equiv \begin{bmatrix} 0 & -i \\ +i & 0 \end{bmatrix} \text{ and } \sigma_z \equiv \begin{bmatrix} 1 & 0 \\ 0 & -1 \end{bmatrix} \qquad (23.20)$$

which are widely used in the spin-related literature.

Making use of the Pauli spin matrices, we could write Eq. (23.19) compactly in the form

$$\Gamma = \frac{\alpha+\beta}{2}\left(\mathbf{I} + \sigma_x\,P_x + \sigma_y\,P_y + \sigma_z\,P_z\right)$$

$$= \frac{\alpha+\beta}{2}\left(\mathbf{I} + \boldsymbol{\sigma}\cdot\mathbf{P}\right). \qquad (23.21)$$

This result applies to the self-energy matrices as well. For example, if

$$\tilde{\Sigma} = -\frac{i}{2}\begin{bmatrix} \alpha & 0 \\ 0 & \beta \end{bmatrix}$$

in the $\pm\,\hat{\mathbf{n}}$ basis, then in the $\pm\,\hat{\mathbf{z}}$ basis it is given by

$$\Sigma = -i\frac{\alpha + \beta}{4}\mathbf{I} - i\frac{\alpha - \beta}{4}\boldsymbol{\sigma}\cdot\hat{\mathbf{n}}$$
$$= -i\frac{\alpha + \beta}{4}(\mathbf{I} + \boldsymbol{\sigma}\cdot\mathbf{P}).$$

23.5   Spin Hamiltonians

Related video lecture available at course website, Unit 4: L4.6.

Now that we have seen how to describe contacts with spin-dependent properties, let us talk briefly about channels with spin-dependent properties.

23.5.1   *Channel with Zeeman splitting*

The commonest example is the Zeeman splitting that causes the energies of the up-spin state to go up by $\mu_{el}B$ and that of the down spin states to go down by $\mu_{el}B$, μ_{el} being the effective magnetic moment of the electron.

If the magnetic field points along $+\,\hat{\mathbf{n}}$, then in the $\pm\,\hat{\mathbf{n}}$ basis the corresponding Hamiltonian should look like

$$\mu_{el}\begin{bmatrix} +B & 0 \\ 0 & -B \end{bmatrix}.$$

Following our discussion in the last section we can write it in the $\pm\,\hat{\mathbf{z}}$ basis as

$$\mathbf{H}_B = \mu_{el}\,\boldsymbol{\sigma}\cdot\mathbf{B}. \tag{23.22}$$

The overall Hamiltonian is obtained by adding this to the spin-independent part multiplied by $\mathbf{I}$. For parabolic dispersion this gives

$$\mathbf{H} = \frac{\hbar^2}{2m}\left(k_x^2 + k_y^2\right)\mathbf{I} + \mu_{el}\boldsymbol{\sigma}\cdot\mathbf{B} \tag{23.23}$$

while for a 2D square lattice we have (see Eq. (17.19))

$$\mathbf{H} = (\varepsilon + 2t\cos(k_x a) + 2t\cos(k_y a))\,\mathbf{I} + \mu_{el}\boldsymbol{\sigma}\cdot\mathbf{B}. \tag{23.24}$$

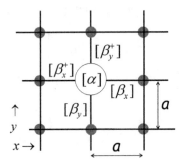

The corresponding parameters for the 2D lattice in Fig. 23.5 (also shown here for convenience) are given simply by

$$\alpha = \varepsilon \mathbf{I} + \mu_{el}\, \boldsymbol{\sigma} \cdot \mathbf{B} \tag{23.25a}$$

$$\boldsymbol{\beta}_x = t\,\mathbf{I}, \;\; \boldsymbol{\beta}_y = t\,\mathbf{I}. \tag{23.25b}$$

Only the on-site parameter $\boldsymbol{\alpha}$ is changed relative to the spin independent channel (Eq. (23.8)).

23.5.2   *Channel with Rashba interaction*

Related video lecture available at course website, Unit 4: L4.5.

A more complicated example is that of the Rashba spin-orbit coupling described by a Hamiltonian of the form

$$\mathbf{H}_R = \eta\,\hat{\mathbf{z}} \cdot (\boldsymbol{\sigma} \times \mathbf{k}) = \eta\,(\sigma_x k_y - \sigma_y k_x) \tag{23.26}$$

whose effect has been observed in 2D surface conduction channels assumed to lie in the x-y plane. This is believed to be a relativistic effect whereby the extremely high atomic scale electric fields (that exist even at equilibrium) are perceived as an effective magnetic field by the electron and the resulting "Zeeman splitting" is described by $\mathbf{H}_R$.

We will not go into the underlying physics of this effect any further here and simply address the question of how to include it in our 2D lattice model. With this in mind we approximate the linear terms with sine functions

$$\mathbf{H}_R = \frac{\eta}{a}\,(\sigma_x \sin(k_y a) - \sigma_y \sin(k_x a)) \tag{23.27}$$

which are written in terms of exponentials:

$$\mathbf{H}_R = \frac{\eta}{2ia}\,\sigma_x(e^{+ik_y a} - e^{-ik_y a}) - \frac{\eta}{2ia}\,\sigma_y(e^{+ik_x a} - e^{-ik_x a}).$$

Clearly $\mathbf{H}_R$ can be described by a Hamiltonian with

$$\boldsymbol{\beta}_x = \frac{i\,\eta}{2a}\,\sigma_y\,,\quad \boldsymbol{\beta}_x^\dagger = -\frac{i\,\eta}{2a}\,\sigma_y$$

$$\boldsymbol{\beta}_y = -\frac{i\,\eta}{2a}\,\sigma_x\,,\quad \boldsymbol{\beta}_y^\dagger = \frac{i\,\eta}{2a}\,\sigma_x$$

in order to ensure that if we write down the dispersion relation for the lattice we will indeed get back the original result in Eq. (23.23). Adding this to the usual spin-independent part from Eq. (23.8) along with any real magnetic field $\mathbf{B}$ we have the overall parameters:

$$\boldsymbol{\alpha} = \varepsilon\mathbf{I} + \mu_{el}\,\boldsymbol{\sigma}\cdot\mathbf{B}$$

$$\boldsymbol{\beta}_x = t\mathbf{I} + \frac{i\,\eta}{2a}\,\sigma_y\,,\quad \boldsymbol{\beta}_x^\dagger = t\mathbf{I} - \frac{i\,\eta}{2a}\,\sigma_y \qquad (23.28)$$

$$\boldsymbol{\beta}_y = t\mathbf{I} - \frac{i\,\eta}{2a}\,\sigma_x\,,\quad \boldsymbol{\beta}_y^\dagger = t\mathbf{I} + \frac{i\,\eta}{2a}\,\sigma_x.$$

23.6 Vectors and Spinors

Related video lecture available at course website, Unit 4: L4.4.

One of the important subtleties that takes some time to get used to is that we represent spin with two complex components, but we visualize it as a rotatable object pointing in some direction, which we have learnt to represent with a vector having three real components. To see the connection between the spinor and the vector, it is instructive to consider the precession of a spin in a magnetic field from both points of view.

Consider the one-level device with $\varepsilon = 0$, and with a magnetic field in the z-direction so that the Schrödinger equation can be written as

$$\frac{d}{dt}\begin{Bmatrix}\psi_u \\ \psi_d\end{Bmatrix} = \frac{\mu_{el}B_z}{i\hbar}\underbrace{\begin{bmatrix}1 & 0 \\ 0 & -1\end{bmatrix}}_{\sigma_z}\begin{Bmatrix}\psi_u \\ \psi_d\end{Bmatrix}. \qquad (23.29)$$

These are two separate differential equations whose solution is easily written down:

$$\psi_u(t) = \psi_u(0)\,e^{-i\omega t/2}$$

$$\psi_d(t) = \psi_d(0)\,e^{+i\omega t/2}$$

$$\text{where}\quad \omega \equiv \frac{2\mu_{el}B_z}{\hbar}. \qquad (23.30)$$

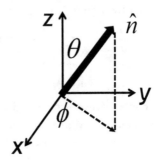

So if the electron starts out at some angle (θ, ϕ) with a wavefunction

$$
\left\{
\begin{aligned}
\psi_u(0) &= \cos\left(\frac{\theta}{2}\right) e^{-i\phi/2} \\
\psi_d(0) &= \sin\left(\frac{\theta}{2}\right) e^{+i\phi/2}
\end{aligned}
\right\}
$$

at $t = 0$, then at a later time it will have a wavefunction given by

$$
\left\{
\begin{aligned}
\psi_u(t) &= \cos\left(\frac{\theta}{2}\right) e^{-i\phi/2}\, e^{-i\omega t/2} \\
\psi_d(t) &= \sin\left(\frac{\theta}{2}\right) e^{+i\phi/2}\, e^{+i\omega t/2}
\end{aligned}
\right\}
$$

which means that the spin will be rotating around the z-axis such that the angle θ remains fixed while the angle ϕ increases linearly with time:

$$
\phi(t) = \phi(0) + \omega t. \tag{23.31}
$$

Making use of Eq. (23.6) for the x, y and z components of the vector $\hat{n}$ we can write

$$
n_x = \sin\theta \cos\phi(t), \quad n_y = \sin\theta \sin\phi(t) \text{ and } n_z = \cos\theta. \tag{23.32}
$$

From Eqs. (23.31) and (23.32) we can show that

$$
\frac{dn_x}{dt} = -\omega\, n_y, \quad \frac{dn_y}{dt} = +\omega\, n_x
$$

which can be written in matrix form

$$
\frac{d}{dt}
\left\{
\begin{array}{c}
n_x \\
n_y \\
n_z
\end{array}
\right\}
= \omega
\underbrace{
\left[
\begin{array}{ccc}
0 & -1 & 0 \\
+1 & 0 & 0 \\
0 & 0 & 0
\end{array}
\right]
}_{\mathbf{R}_z}
\left\{
\begin{array}{c}
n_x \\
n_y \\
n_z
\end{array}
\right\}. \tag{23.33a}
$$

For comparison we have rewritten the Schrödinger equation we started with (see Eq. (23.29)) in terms of the rotation frequency ω:

$$\frac{d}{dt}\left\{\begin{matrix}\psi_u \\ \psi_d\end{matrix}\right\} = \frac{\omega}{2i}\underbrace{\begin{bmatrix} 1 & 0 \\ 0 & -1 \end{bmatrix}}_{\sigma_z}\left\{\begin{matrix}\psi_u \\ \psi_d\end{matrix}\right\}. \tag{23.33b}$$

If we wanted to describe the rotation of an electron due to a **B**-field pointing in the x-direction, it is easy to see how we would modify Eq. (23.33a): Simply interchange the coordinates, $x \to y$, $y \to z$, $z \to x$.

$$\frac{d}{dt}\left\{\begin{matrix}n_x \\ n_y \\ n_z\end{matrix}\right\} = \omega\underbrace{\begin{bmatrix} 0 & 0 & 0 \\ 0 & 0 & -1 \\ 0 & +1 & 0 \end{bmatrix}}_{\mathbf{R}_x}\left\{\begin{matrix}n_x \\ n_y \\ n_z\end{matrix}\right\}$$

and we obtain $\mathbf{R}_x$ in place of $\mathbf{R}_z$. But it is not as clear how to modify Eq. (23.33b). The correct answer is to replace σ_z with σ_x (Eq. (23.20))

$$\frac{d}{dt}\left\{\begin{matrix}\psi_u \\ \psi_d\end{matrix}\right\} = \frac{\omega}{2i}\underbrace{\begin{bmatrix} 0 & 1 \\ 1 & 0 \end{bmatrix}}_{\sigma_x}\left\{\begin{matrix}\psi_u \\ \psi_d\end{matrix}\right\}$$

but the reason is not as obvious.

Equations (23.33a) and (23.33b) both describe the same physics, namely the rotation of a spin about the z-axis due to an applied **B**-field in the z-direction, one in terms of three real components and the other in terms of two complex components.

But what do matrices like **R** in Eq. (23.33a) have in common with matrices like σ in Eq. (23.33b) that makes them "isomorphic" allowing them to describe the same physics? Answer: They obey the same "*commutation relations*". Let me explain.

It is easy to check that the matrices

$$\mathbf{R}_x = \begin{bmatrix} 0 & 0 & 0 \\ 0 & 0 & -1 \\ 0 & +1 & 0 \end{bmatrix}, \ \mathbf{R}_y = \begin{bmatrix} 0 & 0 & +1 \\ 0 & 0 & 0 \\ -1 & 0 & 0 \end{bmatrix}, \ \mathbf{R}_z = \begin{bmatrix} 0 & -1 & 0 \\ +1 & 0 & 0 \\ 0 & 0 & 0 \end{bmatrix}$$

obey the relations

$$\begin{aligned}\mathbf{R}_x\mathbf{R}_y - \mathbf{R}_y\mathbf{R}_x &= \mathbf{R}_z \\ \mathbf{R}_y\mathbf{R}_z - \mathbf{R}_z\mathbf{R}_y &= \mathbf{R}_x \\ \mathbf{R}_z\mathbf{R}_x - \mathbf{R}_x\mathbf{R}_z &= \mathbf{R}_y.\end{aligned} \tag{23.34a}$$

The Pauli spin matrices obey a similar relationship with $\mathbf{R}$ replaced by $\sigma/2i$:

$$\sigma_x\sigma_y - \sigma_y\sigma_x = 2i\,\sigma_z$$
$$\sigma_y\sigma_z - \sigma_z\sigma_y = 2i\,\sigma_x \qquad (23.34b)$$
$$\sigma_z\sigma_x - \sigma_x\sigma_z = 2i\,\sigma_y.$$

The standard textbook introduction to spin starts from these commutation relations and argues that they are a property of the "rotation group". In order to find a mathematical representation with two components for a rotatable object, one must first write down three (2×2) matrices obeying these commutation properties which would allow us to rotate the spinor around each of the three axes respectively.

What are the components of a spinor that points along z? Since rotating it around the z-axis should leave it unchanged, it should be an eigenvector of σ_z that is,

$$\left\{ \begin{matrix} 1 \\ 0 \end{matrix} \right\} \quad \text{or} \quad \left\{ \begin{matrix} 0 \\ 1 \end{matrix} \right\}$$

which indeed represent an upspin and a downspin along z. Similarly if we want the components of a spinor pointing along x, then we should look at the eigenvectors of σ_x, that is,

$$\frac{1}{\sqrt{2}} \left\{ \begin{matrix} +1 \\ +1 \end{matrix} \right\} \quad \text{or} \quad \frac{1}{\sqrt{2}} \left\{ \begin{matrix} +1 \\ -1 \end{matrix} \right\}$$

which represent up and down spin along $+x$. If we consider a spinor pointing along an arbitrary direction described by a unit vector $\hat{\mathbf{n}}$ (see Eq. (23.6)) and wish to know what its components are, we should look for the eigenvectors of

$$\boldsymbol{\sigma} \cdot \hat{\mathbf{n}} = \sigma_x \sin\theta \cos\phi + \sigma_y \sin\theta \sin\phi + \sigma_z \cos\theta$$
$$= \begin{bmatrix} \cos\theta & \sin\theta\, e^{-i\phi} \\ \sin\theta\, e^{+i\phi} & -\cos\theta \end{bmatrix} \qquad (23.35a)$$

which can be written as (c and s defined in Eq. (23.7))

$$\left\{ \begin{matrix} c \\ s \end{matrix} \right\} \quad \text{and} \quad \left\{ \begin{matrix} -s^* \\ c^* \end{matrix} \right\}. \qquad (23.35b)$$

In short, the rigorous approach to finding the spinor representation is to first determine a set of three matrices with the correct commutation relations and then look at their eigenvectors. Instead in this chapter, I adopted a reverse approach stating the spinor components at the outset and then obtaining the matrices through basis transformations.

23.7 Spin Precession

We have already discussed how to write $\mathbf{H}$ and $\mathbf{\Sigma}$ including non-trivial spin-dependent effects and we could set up numerical models to calculate the electron density $\mathbf{G}^n$, or the density of states $\mathbf{A}$, or the current using the standard NEGF equations from Chapter 18. Consider for example, the non-local spin potential measurement we started this chapter with (see Fig. 23.8).

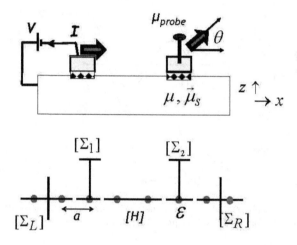

Fig. 23.8 Spin potential measurement can be modeled with a 1D channel Hamiltonian having four contacts, two of which are magnetic described by $\mathbf{\Sigma}_1$, $\mathbf{\Sigma}_2$.

Figure 23.9 shows the result obtained from the numerical model which supports the basic result stated in Eq. (23.5). The measured voltage oscillates as a function of the angle of magnetization of the voltage probe. It has a constant part independent of the angle and an oscillatory component proportional to the polarization P of the voltage probe which can be understood in terms of Eq. (23.5) stated at the beginning of this chapter.

I am not sure if the experiment shown in Fig. 23.8 has been done, but what has been done is to keep both magnets fixed and rotate the electron spin inside the channel.

How do we rotate the spin? One method that has been widely used is an external magnetic field $\mathbf{B}$ which causes the spin direction to precess around the magnetic field as we discussed in Section 23.6 with an angular

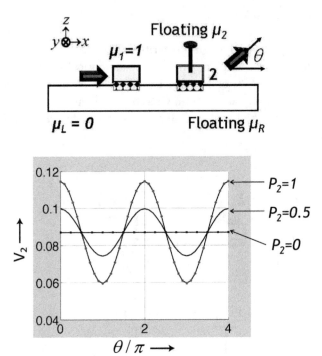

Fig. 23.9 Voltage probe signal as the magnetization of the probe is rotated calculated from NEGF model. For MATLAB script, see Appendix H.4.1.

frequency given by

$$\omega \equiv \frac{2\mu_{el}B_z}{\hbar} \quad \text{(same as Eq. (23.30))}. \quad (23.36)$$

This means that the spin voltage at the point where the probe is connected will rotate by an angle $\omega\tau$ where τ is the time it takes for the electron to travel from the point of injection to the voltage probe. Writing $\tau = L/v$, we have from Eq. (23.5) using Eq. (23.30)

$$\mu_P = \mu + P_2\,\mu_s \cos\left(\frac{2\mu_{el}L}{\hbar v}B_z\right). \quad (23.37)$$

One would expect to see an oscillatory signal as a function of the magnetic field. However, one is usually not dealing with ballistic transport, and there is a large spread in the time τ spent by an electron between injection and detection so that the average value of this signal over all τ is essentially zero. What is typically observed is not an oscillatory signal as a function of the **B**-field but a reduction in the signal from $P\mu_s$ down to zero, which is referred to as the *Hanle signal*. However, Hanle signals showing several oscillations have also been observed, but this requires that the spread in τ be much less than its mean value (see for example, Huang *et al.*, 2007).

Another possible approach to rotating electron spins is to use the Rashba effect in materials with strong spin-orbit coupling. In many semiconductors, it is now well established that a surface electric field along z (Fig. 23.9) leads to an effective magnetic field that depends on the electron momentum. This can be seen by comparing the Hamiltonians for the **B**-field (Eq. (23.22))

$$\mathbf{H}_B = \mu_{el}\,\boldsymbol{\sigma}.\mathbf{B}$$

with that for the Rashba interaction (23.26) which can be rewritten as

$$\mathbf{H}_R = \eta\,\hat{\mathbf{z}}\cdot(\boldsymbol{\sigma}\times\mathbf{k}) \quad\rightarrow\quad \eta\,\boldsymbol{\sigma}\cdot(\hat{\mathbf{z}}\times\mathbf{k})$$

suggesting that the effective **B**-field due to the Rashba interaction is given by

$$\mu_{el}\mathbf{B}_{eff} = \eta\,\hat{\mathbf{z}}\times\mathbf{k} \tag{23.38}$$

so that from Eq. (23.37) we expect an oscillatory signal of the form

$$\mu_P = \mu + P_2\mu_s \cos\left(\frac{2\eta kL}{\hbar\nu}\right) \tag{23.39}$$

with a period $\Delta\eta$ defined by

$$\frac{2kL}{\hbar\nu}\Delta\eta = 2\pi \quad\rightarrow\quad \Delta\eta = \frac{2\pi a t_0}{kL}\sin(ka).$$

This is in approximate agreement with the numerical result obtained from the NEGF method (Fig. 23.10) using an energy E corresponding to $ka = \pi/3$, and a distance of about $L = 40\,a$ between the injector and the detector.

In the structure shown in Fig. 23.10 the electrons traveling along $+x$ should feel an effective **B**-field along y. Since the injected spins have a spin voltage $\boldsymbol{\mu}_s$ pointing along the source and drain magnets (x) it should be rotated. Note that the oscillation should not be observed if the source and drain magnets point along y rather than along x.

This phenomenon of voltage-controlled spin precession in 2D conductors with high spin-orbit coupling was predicted in 1990 (Datta & Das, 1990) and has now been experimentally established (Koo *et al.*, 2009, Wunderlich *et al.*, 2010, Choi *et al.*, 2015).

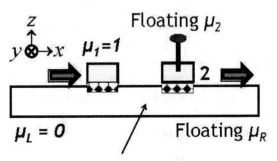

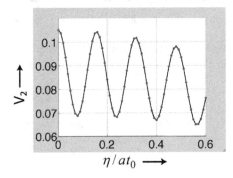

Fig. 23.10 In materials with a large Rashba coefficient, a gate voltage should lead to an oscillatory output, if the source and drain magnets point along x, but not if they point along z. For MATLAB script, see Appendix H.4.1.

23.8 Spin-charge Coupling

Spin precession is a delicate effect requiring a degree of spin coherence that is usually available only at low temperatures. That is why voltage-controlled spin precession seems to have little potential as an electronic device although it is the basis for many proposals for spin transistors. Interestingly, the same basic spin-orbit interaction leads to the phenomenon of spin-momentum locking which gives rise to robust room temperature effects that are finding extensive applications as we discussed in Part A (see Section 12.2.3, Part A).

The phenomenon can be understood by noting that due to the Rashba interaction (see Eq. (23.26)) an electron in a state $\mathbf{k}$ feels an effective magnetic field that can be written as

$$\mu_B \mathbf{B}_{eff} = \eta\, \hat{\mathbf{z}} \times \mathbf{k}. \tag{23.40}$$

This leads to an $E(\mathbf{k})$ relation of the form

$$E(\mathbf{k}) \;=\; \frac{\hbar^2 k^2}{2m} \pm \mu_B B_{eff} \;=\; \frac{\hbar^2 k^2}{2m} \pm \eta k \tag{23.41}$$

where the positive and negative signs correspond to spins being parallel or anti-parallel to $\mathbf{B}_{eff}$.

We can turn this $E(\mathbf{k})$ relation around to note that for a given E, there are two distinct allowed values of k given by

$$k_{f2} = \sqrt{k_F^2 + k_0^2} \,+\, k_0 \quad\text{and}\quad k_{f1} = \sqrt{k_F^2 + k_0^2} \,-\, k_0 \tag{23.42a}$$

$$\text{where}\quad k_F = \frac{2mE}{\hbar^2}\ ,\quad k_0 = \frac{\eta m}{\hbar^2}. \tag{23.42b}$$

In other words, for a given energy E, there are two Fermi circles with radii k_{f1} and k_{f2} (see Fig. 23.11a), corresponding to states with spins parallel and anti-parallel to $\mathbf{B}_{eff}$ respectively.

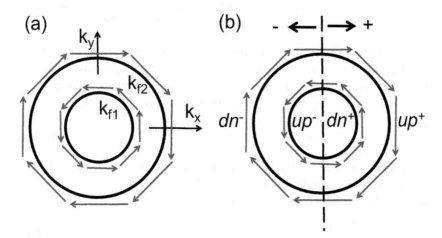

Fig. 23.11 (a) Surface states in materials with high spin-orbit coupling have two different Fermi circles with radii k_{f1} and k_{f2}. Arrows show the direction of the spin associated with these eigenstates which is parallel and antiparallel to $\mathbf{B}_{eff}$ respectively. (b) The states are divided into '+' and '−' depending on the sign of the group velocity: Each has two branches labeled 'up' and 'dn' depending on the sign of the y-component of the spin. Overall we have four groups $\mathrm{up}^+, \mathrm{dn}^-, \mathrm{up}^-, \mathrm{dn}^+$.

In discussing current flow in ordinary conductors we found it useful to divide all states into positive and negative moving states, '+' and '−' and assigning them different electrochemical potentials μ^+ and μ^- (see Chapters 8–11, Part A). In these 2D conductors with spin-orbit coupling we can do the same as shown in Fig. 23.11b, but each group now has two branches corresponding to the two Fermi circles. We label these branches depending on the sign of the y-component of the spin, up: $\hat{\mathbf{y}} \cdot \hat{\mathbf{s}} < 0$ and dn: $\hat{\mathbf{y}} \cdot \hat{\mathbf{s}} > 0$. Overall we have four groups $\text{up}^+, \text{dn}^-, \text{up}^-, \text{dn}^+$.

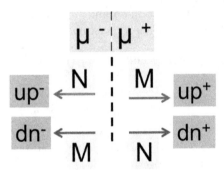

Fig. 23.12   Surface states in materials with high spin-orbit coupling have equal number of modes M for up^+, dn^-, but a different number of modes N for up^-, dn^+.

Two of these groups up^+ and dn^- correspond to the larger Fermi circle of radius k_{f2} and have a number of modes M that is greater than the number of modes N for the other two up^- and dn^+ which correspond to the smaller Fermi circle of radius k_{f1} (W: Width of conductor):

$$M = \frac{k_{f2}W}{\pi} \quad \text{and} \quad N = \frac{k_{f1}W}{\pi}. \tag{23.43}$$

As we discussed in Part A (Chapters 8–12) whenever a current I flows in *any* material, there is a separation in the electrochemical potentials μ^+ and μ^- (Eq. (12.24), G_B: Ballistic conductance)

$$I = G_B \frac{\mu^+ - \mu^-}{q}. \tag{23.44}$$

But in the materials with spin-orbit coupling that we are discussing, this separation in μ^+ and μ^- manifests itself as a difference in the electrochemical potentials for μ^{up} and μ^{dn}:

$$\mu^{up} = \frac{M\mu^+ + N\mu^-}{M + N} \quad \text{and} \quad \mu^{dn} = \frac{N\mu^+ + M\mu^-}{M + N} \tag{23.45}$$

giving rise to a spin potential (see Eq. (23.1)) in the y-direction

$$\mu_{sy} = \frac{\mu^{up} - \mu^{dn}}{2} = p\frac{\mu^+ - \mu^-}{2} \tag{23.46}$$

where we have defined the channel polarization as

$$p \sim \frac{M - N}{M + N}. \tag{23.47}$$

In Eq. (23.47) we are not using the equality sign since we have glossed over a "little" detail involving the fact that the y-component of the spins in each of the groups up$^{\pm}$ and dn$^{\pm}$ has a distribution (see Fig. 23.11), which on averaging gives rise to a numerical factor.

Making use of Eqs. (23.44) and (23.46) we can rewrite the y-component of the spin potential in terms of the current I:

$$\mu_{sy} = \frac{q}{2G_B}pI. \tag{23.48}$$

Note that the channel polarization "p" appearing in Eq. (23.48) is a channel property that determines the intrinsic spin potential appearing in the channel. It is completely different from the probe polarization "P" defined in Eq. (23.3) which is a magnet property that comes into the picture only when we use a y-directed magnetic probe to measure the intrinsic spin potential μ_{sy} induced in the channel by the flow of current (I).

This is a remarkable result that shows a new way of generating spin potentials. We started this chapter with spin valves that generate spin potentials through the spin-dependent interface resistance of magnetic contacts. By contrast Eq. (23.48) tells us that a spin voltage can be generated in channels with spin-momentum locking simply by the flow of current without the need for magnetic contacts, arising from the difference between M and N. The spin voltage can be measured with magnetic contacts.

Alternatively we could reverse the voltage and current terminals and invoke reciprocity (Section 10.3.3) to argue that a current injected through a magnetic contact will generate a charge voltage at the ordinary contacts.

This is our view of the Rashba-Edelstein (RE) effect and its inverse which have been observed in a wide variety of materials like topological insulators and narrow gap semiconductors. Similar effects are also observed in heavy metals where it is called the (inverse) spin Hall effect (SHE) and is often associated with bulk scattering mechanisms, but there is some evidence that it could also involve the surface mechanism described here.

23.9   Superconducting Contacts

Let me end this chapter (which is too long already) with a brief note about an area of great current interest, that can be addressed using the NEGF methods described in this book. This area involves electrical conduction in devices with normal channel materials like the ones we have been talking about, but one or both contacts (source / drain) are superconducting.

It takes considerable discussion to clarify the conceptual basis for handling superconducting contacts and the associated phenomenon of Andreev scattering, but the key point is that the "quasi-particles" are not described by the Hamiltonian $\mathbf{H}$ that we have been discussing but by the *Bogoliubov-deGennes* (BdG) Hamiltonian, $\mathbf{H}_{BdG}$.

If the channel has no spin-related properties and can be described by a spinless Hamiltonian $\mathbf{H}$, then structures with superconducting contacts can be described by a $(2 \times 2)\,\mathbf{H}_{BdG}$ (see for example, deGennes, 1968)

$$\mathbf{H}_{\mathrm{BdG}} = \begin{bmatrix} \mathbf{H} - \mu\mathbf{I} & \Delta \\ \Delta^* & -(\mathbf{H}^* - \mu\mathbf{I}) \end{bmatrix} \tag{23.49}$$

$\mathbf{I}$ being the identity matrix of the same size as $\mathbf{H}$, and Δ being the order parameter of the superconductor, which is set to zero for the normal contact(s) and the channel. The electrochemical potential μ is set to zero for the superconducting contact.

Using $\mathbf{H}_{BdG}$ we can find the Σ's for the contacts in the same way and proceed to use the same NEGF equations as summarized in Appendix G. Interpreting the results, however, will require an appreciation of the conceptual basis underlying the BdG equation (see Datta and Bagwell, 1999).

Note that the electrochemical potential μ is assumed zero in the superconducting contact. But this is not possible if there are multiple superconducting contacts held at different potentials. The calculation is then more involved because solutions at different energies get coupled together (see Samanta and Datta, 1997).

Finally I should mention that there is a lot of current interest in channel materials whose Hamiltonian includes spin-orbit coupling and other spin-related properties, and this requires a $(4\times4)\,\mathbf{H}_{BdG}$ (see for example, Pikulin *et al.*, 2012; San-Jose *et al.*, 2013).

Chapter 24

Quantum to Classical

Related video lecture available at course website, Unit 4: L4.1.

In this book (Part B) we have introduced the formalism of quantum transport, while Part A was based on the semiclassical view that treats electrons primarily as particles, invoking its wave nature only to obtain the density of states, $D(E)$ and the number of modes, $M(E)$. The distinction between quantum and classical viewpoints is a much broader topic of general interest and we believe that spin transport provides a natural framework for understanding and exploring at least some aspects of it. Let me try to elaborate with a few thoughts along these lines.

24.1 Matrix Electron Density

Related video lecture available at course website, Unit 4: L4.7.

Earlier we talked about the connection between vector $\hat{n}$ along which a spin points and the wavefunction ψ representing it. To relate Eq. (23.5) to the NEGF method we need to consider quantities like $\mathbf{G}^n \sim \psi\psi^\dagger$ rather than the wavefunction ψ, since the NEGF is formulated in terms of $\mathbf{G}^n$. Besides it is $\mathbf{G}^n$ and not ψ that is observable and can be related to experiment.

We have often referred to $\mathbf{G}^n$ as the matrix electron density whose diagonal elements tell us the number of electrons at a point. With spin included, $\mathbf{G}^n$ at a point is a (2×2) matrix and the elements of this matrix tell us the number of electrons N or the net number of spins $\mathbf{S}$.

To see this consider an electron pointing in some direction $\hat{n}$ represented by a spinor wavefunction of the form (see Eq. (23.7))

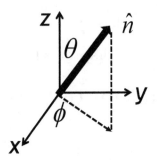

$$\psi = \left\{ \begin{array}{l} \cos\left(\dfrac{\theta}{2}\right) e^{-i\phi/2} \equiv c \\[2mm] \sin\left(\dfrac{\theta}{2}\right) e^{+i\phi/2} \equiv s \end{array} \right\}$$

so that the corresponding (2×2) electron density $\mathbf{G}^n$ is given by

$$\psi\psi^\dagger = \left\{ \begin{array}{c} c \\ s \end{array} \right\} \left\{ c^* \ \ s^* \right\} = \left[\begin{array}{cc} cc^* & cs^* \\ sc^* & ss^* \end{array} \right].$$

Making use of Eq. (23.6) and Eq. (23.17) we have

$$\psi\psi^\dagger = \frac{1}{2} \left[\begin{array}{cc} 1 + n_z & n_x - i\,n_y \\ n_x + i\,n_y & 1 - n_z \end{array} \right] = \frac{1}{2}\, [\mathbf{I} + \boldsymbol{\sigma} \cdot \hat{\mathbf{n}}].$$

For a collection of N electrons we can add up all their individual contributions to $\psi\psi^\dagger$ to obtain the net $\mathbf{G}^n$ given by

$$\frac{\mathbf{G}^n}{2\pi} = \frac{1}{2} \left[\begin{array}{cc} N + S_z & S_x - i\,S_y \\ S_x + iS_y & N - S_z \end{array} \right] \tag{24.1}$$

$$= \frac{1}{2}\, (N\mathbf{I} + \boldsymbol{\sigma} \cdot \mathbf{S}).$$

Given a $\mathbf{G}^n$ we can extract these quantities from the relations

$$N = \frac{1}{2\pi}\, \text{Trace}\,[\mathbf{G}^n], \quad \mathbf{S} = \frac{1}{2\pi}\, \text{Trace}\,[\boldsymbol{\sigma}\,\mathbf{G}^n] \tag{24.2}$$

which follow from Eq. (24.1) if we make use of the fact that all three matrices (Eq. (23.20)) have zero trace, along with the following properties of the Pauli spin matrices that are easily verified.

$$\sigma_x^2 = \sigma_y^2 = \sigma_z^2 = \mathbf{I} \tag{24.3a}$$

$$\sigma_x\sigma_y = -\sigma_y\sigma_x = i\sigma_z \qquad (24.3b)$$

$$\sigma_y\sigma_z = -\sigma_z\sigma_y = i\sigma_x \qquad (24.3c)$$

$$\sigma_z\sigma_x = -\sigma_x\sigma_z = i\sigma_y. \qquad (24.3d)$$

In summary, all the information contained in the (2×2) Hermitian matrix $\mathbf{G}_n$ can be expressed in terms of four real quantities consisting of a scalar N and the three components of a vector $\mathbf{S}$ which can be extracted using Eq. (24.2).

If the transverse components of spin are negligible then we can describe the physics in terms of N and S_z only. We could interpret the non-zero components on the diagonal

$$(N + S_z) \text{ as number of up electrons}$$

$$(N - S_z) \text{ as number of down electrons}$$

(per unit energy) and then write semiclassical equations for the two types of electrons.

Consider for example, the experiment that we discussed in Chapter 23. An input magnet injects spins into the channel which produce a voltage on the output magnet given by

$$\mu_P = \mu + \mathbf{P} \cdot \boldsymbol{\mu}_s. \qquad (24.4)$$

When can we understand these measurements just in terms of up and down spins? One possibility is that the magnets are all collinear and there is no spin-orbit coupling so that we are restricted to angles θ that are multiples of 180 degrees (Fig. 24.1). Another possibility is that various spin dephasing

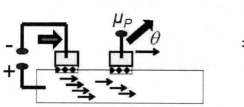
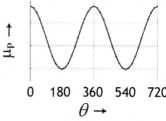

Fig. 24.1 An input magnet injects spins into the channel which produce a voltage on the output magnet that depends on the cosine of the angle between the two magnets.

processes are strong enough to reduce transverse components of spin to negligible proportions. And if the z-components are reduced too, then we would not have to worry about spin at all.

What if we had collinear magnets but they point not along z, but along x? Now the $\mathbf{G}^n$ matrix is not diagonal

$$\frac{\mathbf{G^n}}{2\pi} \rightarrow \begin{bmatrix} N & S_x \\ S_x & N \end{bmatrix}$$

and it might appear that a semiclassical description is not possible. The trick is to choose the coordinates or more generally the "basis" correctly. What we should do is to select a basis in which up and down point along $+x$ and $-x$ respectively so that in this basis $\mathbf{G}^n$ is diagonal

$$\frac{\mathbf{G}^n}{2\pi} \rightarrow \begin{bmatrix} N + S_{up} & 0 \\ 0 & N + S_{dn} \end{bmatrix}.$$

In a word, we should simply call the direction of the magnet z instead of x!

This sounds like a trivial observation, but it represents a general truth. Later in Section 24.4 we will discuss the concept of *pseudo-spins*, or diverse quantum objects that can be viewed as spins. Such pseudo-spins too can also be visualized in classical terms, if we use a basis in which off-diagonal elements play a negligible role. More generally, the wave function $\boldsymbol{\psi}$ for any quantum object can be viewed as a *giant spin* with a large number of components, and classical visualizations are accurate when the off-diagonal elements of $\boldsymbol{\psi\psi}^\dagger$ play a negligible role.

24.2   Matrix Potential

Related video lecture available at course website, Unit 4: L4.8.

To get a little more insight, let us now consider the *matrix* version of the classical model for a voltage probe that we used earlier (see Fig. 23.2) to obtain the scalar version of the result stated at the beginning of Chapter 23:

$$\mu_\mathrm{P} = \mu + \mathbf{P} \cdot \boldsymbol{\mu}_s \quad \text{(Same as Eq. (23.5)).}$$

Now we can obtain the general vector version by starting from the NEGF model for a probe (Fig. 24.2) with the current given by (see Eq. (18.4))

$$I \sim \text{Trace}\left(\boldsymbol{\Gamma}\left[f_\mathrm{P}\mathbf{A} - \mathbf{G}^n\right]\right)$$

so that for zero probe current we must have

$$f_\mathrm{P} = \frac{\text{Trace}\left[\boldsymbol{\Gamma}\,\mathbf{G}^n\right]}{\text{Trace}\left[\mathbf{A}\right]}.$$

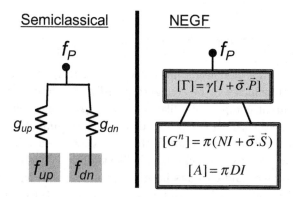

Fig. 24.2   Model for a probe connected to a channel.

Making use of Eq. (24.1) for $\mathbf{G}^n$, assuming that the density of states (D) is spin-independent

$$\frac{\mathbf{A}}{2\pi} = \frac{D}{2}\mathbf{I}$$

writing the probe coupling in the form (see Eq. (23.21))

$$\mathbf{\Gamma} = \gamma\,(\mathbf{I} + \boldsymbol{\sigma}\cdot\mathbf{P})$$

and noting that the Pauli matrices all have zero trace, we obtain

$$f_{\mathrm{P}} = \mathrm{Trace}\left([\mathbf{I} + \boldsymbol{\sigma}\cdot\mathbf{P}]\left[\frac{N}{D}\mathbf{I} + \boldsymbol{\sigma}\cdot\frac{\mathbf{S}}{D}\right]\right). \qquad (24.5)$$

Once again there is an identity that can be used to simplify this expression: For any two vectors $\mathbf{P}$ and $\mathbf{B}$, it is straightforward (but takes some algebra) to verify that

$$[\boldsymbol{\sigma}\cdot\mathbf{P}]\,[\boldsymbol{\sigma}\cdot\mathbf{B}] = (\mathbf{P}\cdot\mathbf{B})\,\mathbf{I} + i\boldsymbol{\sigma}\cdot[\mathbf{P}\times\mathbf{B}],$$

$$\rightarrow [\mathbf{I} + \boldsymbol{\sigma}\cdot\mathbf{P}]\,[b\mathbf{I} + \boldsymbol{\sigma}\cdot\mathbf{B}] = (b + \mathbf{P}\cdot\mathbf{B})\,\mathbf{I} + \boldsymbol{\sigma}\cdot[\mathbf{P} + \mathbf{B} + i\mathbf{P}\times\mathbf{B}]. \qquad (24.6)$$

Making use of this identity and noting once again that the Pauli matrices have zero trace, we can write from Eq. (24.5)

$$f_{\mathrm{P}} = \frac{N}{D} + \mathbf{P}\cdot\frac{\mathbf{S}}{D} \equiv f + \mathbf{P}\cdot\mathbf{f}_s \qquad (24.7)$$

in terms of the charge occupation f and the spin occupation $\mathbf{f}_s$. We can translate these occupations into potentials μ and μ_s, using the linear relation between the two for small bias (see Eq. (2.11)) to obtain the relation stated earlier (see Eq. (23.5)).

24.3 Spin Circuits

Related video lecture available at course website, Unit 4: L4.9.

The NEGF method leads naturally to a (2×2) complex matrix $\mathbf{G}^n$ at each point, which can be straightforwardly translated into four physically transparent components, like N, $\mathbf{S}$ or μ, $\boldsymbol{\mu}_S$, which represent a four-component *voltage*. We could also define four-component currents I, $\mathbf{I}_S$ with 4×4 *conductance matrices* $\mathbf{G}$ that connect voltages to currents. The resulting *spin circuits* look superficially like traditional circuits, except that the nodal voltages and currents have four components each. Similarly one could use electron densities with four components N, $\mathbf{S}$ to write a differential equation superficially similar to the diffusion equation.

The question one could ask is whether these four-component formulations are equivalent to the NEGF method discussed here. The answer is that they capture a subset of the effects contained in the NEGF and there are many problems where this subset may be adequate. Let me explain.

At the beginning of Chapter 23, I mentioned that including spin increases the size of the matrices by a factor of two since every point along z effectively becomes two points, an up and a down. So if there are three points in our channel, the matrix $\mathbf{G}^n$ will be of size (6×6).

Instead of using the entire $\mathbf{G}^n$ matrix, as we do in the full NEGF method we could use just the (2×2) diagonal blocks of this matrix, representing each block with four components. What we will clearly miss is the information contained in the off-diagonal elements between two spatial elements which as we saw in Chapter 19 gives rise to quantum interference effects. But we may not be missing much, since as we discussed, dephasing processes often destroy these interference effects anyway.

Spin information is usually more robust. While phase relaxation times are often sub-picosecond, spin relaxation times are much longer, in nanoseconds. And so it is important to retain the information in the (2×2) diagonal blocks, even if we are throwing away the rest. This four-component spin diffusion approach will reduce to standard two-component spin diffusion (the celebrated Valet-Fert equation) if spin dephasing processes were strong enough to destroy the transverse components $S_{x,y}$ of the spin.

Formally we could do that starting from the NEGF method by defining a suitable **D**-matrix of the type discussed in Chapter 18 relating the inscattering to the electron density ($\times$ denotes element by element multiplication)

$$\mathbf{\Sigma}^{in} = \mathbf{D} \times \mathbf{G}^n \quad \text{(Same as Eq. (18.35b))}.$$

The dephasing process can be viewed as extraction of the electron from a state described by $\mathbf{G}^n$ and reinjecting it in a state described by $\mathbf{D} \times \mathbf{G}^n$. We introduced two models A and B with **D** defined by Eqs. (18.37) and (18.38) respectively. Model A was equivalent to multiplying $\mathbf{G}^n$ by a constant so that the electron was reinjected in exactly the same state that it was extracted in, causing no loss of momentum, while Model B threw away the off-diagonal elements causing loss of momentum as we saw in the numerical example in Fig. 19.7. We could define a **Model C** having a **D**-matrix that retains spin information while destroying momentum:

		$1up$	$1dn$	$2up$	$2dn$	$3up$	$3dn$
	$1up$	1	1	0	0	0	0
	$1dn$	1	1	0	0	0	0
$\dfrac{\mathbf{D}}{D_0} =$	$2up$	0	0	1	1	0	0
	$2dn$	0	0	1	1	0	0
	$3up$	0	0	0	0	1	1
	$3dn$	0	0	0	0	1	1

$$. \quad (24.8)$$

One could view this as Model B-like with respect to the lattice, but Model A-like with respect to spin. We could rewrite the NEGF equation

$$\mathbf{G}^n = \mathbf{G}^R \, \mathbf{\Sigma}^{in} \, \mathbf{G}^A$$

$$\text{as} \quad [G^n]_{i,i} = \sum_j [G^R]_{i,j} \, [\Sigma^{in}]_{j,j} \, [G^A]_{j,i}$$

$$= D_0 \sum_j [G^R]_{i,j} \, [G^n]_{j,j} \, [G^A]_{j,i} \quad (24.9)$$

where the indices i, j refer to lattice points and we have made use of the fact that in our Model C, $\mathbf{\Sigma}^{in}$ is diagonal as far as the lattice is concerned.

We have seen earlier that at any point on the lattice the 2×2 matrix $\mathbf{G}^n$ can be expressed in terms of four components, namely N and $\mathbf{S}$ so that with a little algebra we could rewrite Eq. (24.9) in the form

$$\left\{\begin{array}{c} N \\ S_x \\ S_y \\ S_z \end{array}\right\}_i = \sum_j \left[\begin{array}{c} 4 \times 4 \\ \text{``Hopping''} \\ \text{Matrix} \end{array}\right]_{i,j} \left\{\begin{array}{c} N \\ S_x \\ S_y \\ S_z \end{array}\right\}_j \qquad (24.10)$$

where the (4×4) matrix could be viewed as describing the probability of the $(N, \mathbf{S})$ at a point j hopping to a point i in one time step. Indeed the (1×1) version of Eq. (24.10) resembles the standard description of Brownian motion on a lattice that leads to the drift-diffusion equation.

Spin diffusion equations based on alternative approaches like the Kubo formalism have been discussed in the past (see for example, Burkov *et al.*, 2004). The main point I want to convey is that NEGF-based approaches can also be used to justify and benchmark spin diffusion models which could well capture the essential physics and provide insights that a purely numerical calculation misses.

24.4 Pseudo-spin

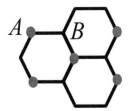

One of the nice things about the formalism of spin matrices (Chapter 23) is that it goes way beyond spins; it applies to any two-component complex quantity. For example in Chapters 17 and 18 we talked about the graphene lattice where the unit cell has an "A" atom (on the lattice sites marked with a red circle) and a "B" atom (on the unmarked lattice sites).

The wavefunction in a unit cell is described by a two component complex quantity:

$$\boldsymbol{\psi} = \left\{\begin{array}{c} \psi_A \\ \psi_B \end{array}\right\}$$

and we could look at the corresponding $\mathbf{G}^n$ and use our old relation from Eq. (24.1) to define a pseudo-spin

$$\frac{\mathbf{G}^n}{2\pi} = \begin{bmatrix} \psi_A \psi_A^* & \psi_A \psi_B^* \\ \psi_B \psi_A^* & \psi_B \psi_B^* \end{bmatrix} \rightarrow \begin{bmatrix} N + S_z & S_x - iS_y \\ S_x + iS_y & N - S_z \end{bmatrix}.$$

This has nothing to do with the real spin, just that they share the same mathematical framework. Once you have mastered the framework, there is no need to re-learn it, you can focus on the physics. In the literature on graphene, there are many references to pseudo spin and what direction it points in.

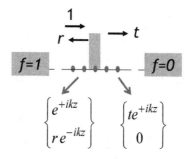

Let me point out a less familiar example of pseudospin involving an example we have already discussed. In Chapter 19, we discussed the potential variation across a single scatterer with transmission equal to T (Fig. 19.7). Let us just look at the diagonal elements of $\mathbf{G}^n$ for the same problem. There are oscillations on the left of the barrier with a constatnt density on the right. The reason Fig. 19.7 shows oscillations on the right as well is that we were looking at the occupation obtained from $G^n(j,j)/A(j,j)$ and $\mathbf{A}$ has oscillations on the right. But let us not worry about that.

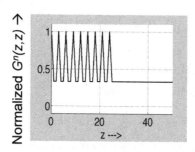

Let us see how we can use pseudospins to understand the spatial variation of the diagonal elements of $\mathbf{G}^n$ by viewing positive and negative going states as the up and down components of a pseudospin. The pseudospinor wavefunction on the left and right of the barrier have the form

$$\begin{array}{cc} Left & Right \end{array}$$

$$\psi \rightarrow \left\{ \begin{array}{c} e^{+ikz} \\ re^{-ikz} \end{array} \right\}, \left\{ \begin{array}{c} te^{+ikz} \\ 0 \end{array} \right\}$$

$$\begin{array}{cc} Left & Right \end{array}$$

$$\psi\psi^{\dagger} \rightarrow \left[\begin{array}{cc} 1 & r^{*}e^{+i2kz} \\ re^{-2ikz} & rr^{*} \end{array} \right], \left[\begin{array}{cc} tt^{*} & 0 \\ 0 & 0 \end{array} \right]$$

$$\rightarrow \left[\begin{array}{cc} N + S_{z} & S_{x} - iS_{y} \\ S_{x} + iS_{y} & N - S_{z} \end{array} \right].$$

This suggests that the pseudospins to the left of the barrier are described by (assuming r and t are real)

$$\begin{array}{cc} Left & Right \end{array}$$

$$\begin{array}{cc} N = (1+r^{2})/2 & N = t^{2}/2 \\ S_{z} = (1-r^{2})/2 & S_{z} = t^{2}/2 \\ S_{x} = r\cos(2kz) & S_{x} = 0 \\ S_{y} = -r\sin(2kz) & S_{y} = 0. \end{array}$$

In other words, on the left of the barrier, the pseudospin is rotating round and round in the x-y plane. When we plot $G^{n}(z,z)$, we are looking at the sum of the two pseudospin components and squaring the sum, which amounts to

$$\text{Trace}\left(\{1 \ 1\} \psi\psi^{\dagger} \left\{ \begin{array}{c} 1 \\ 1 \end{array} \right\} \right) = \text{Trace}\left(\left[\begin{array}{cc} 1 & 1 \\ 1 & 1 \end{array} \right] \psi\psi^{\dagger} \right).$$

In effect we are using a pseudomagnet with $\mathbf{\Gamma} = \left[\begin{array}{cc} 1 & 1 \\ 1 & 1 \end{array} \right]$ which corresponds to one polarized 100% along x. So from Eq. (24.4), the measured potential should be proportional to

$$\begin{array}{cc} Left & Right \end{array}$$

$$N + \hat{\mathbf{x}} \cdot \mathbf{S} \rightarrow \quad \dfrac{1+r^{2}}{2} + r\cos(2kz) \qquad \dfrac{t^{2}}{2}$$

which describes the numerical results quite well.

This is a relatively familiar problem where the concept of pseudospin probably does not add much to our undergraduate understanding of one-dimensional standing waves. The purpose was really to add our understanding of pseudospins!

24.5 Quantum Information

Now that we have seen how "spins" appear everywhere, let us talk briefly about the information content of a single spin which as we discussed in Chapter 16 is related to the thermodynamic entropy.

24.5.1 *Quantum entropy*

We talked about the entropy of two examples of a collection of N spins obtained from the expression

$$\frac{S}{k} = -\sum_i p_i \ln(p_i). \qquad (24.11)$$

From a quantum mechanical point of view we could write the wavefunction of a single spin in collection A as

$$\psi = \begin{Bmatrix} 1 \\ 0 \end{Bmatrix} \quad \rightarrow \quad \psi\psi^\dagger = \begin{bmatrix} 1 & 0 \\ 0 & 0 \end{bmatrix}$$

and interpret the diagonal elements of $\psi\psi^\dagger$ (1 and 0) as the p_i's to use in Eq. (24.11). Writing $\psi\psi^\dagger$ for a spin in collection B requires us to take a sum of two equally likely possibilities:

$$\psi\psi^\dagger = 0.5 \begin{bmatrix} 1 & 0 \\ 0 & 0 \end{bmatrix} + 0.5 \begin{bmatrix} 0 & 0 \\ 0 & 1 \end{bmatrix} = \begin{bmatrix} 0.5 & 0 \\ 0 & 0.5 \end{bmatrix}.$$

Once again we can interpret the diagonal elements of $\psi\psi^\dagger$ (both 0.5) as the p_i's to use in Eq. (24.11) and get our semiclassical answers. What if we have collection C, which looks just like collection A, but the spins all pointing along x and not z.

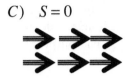

C) $S = 0$

We then have

$$\psi = \frac{1}{\sqrt{2}} \begin{Bmatrix} 1 \\ 1 \end{Bmatrix} \quad \rightarrow \quad \psi\psi^\dagger = \begin{bmatrix} 0.5 & 0.5 \\ 0.5 & 0.5 \end{bmatrix}.$$

If we just took the diagonal elements of $\psi\psi^\dagger$ (both 0.5) we obtain the same answer as we got for collection B which is obviously wrong. A collection with all spins pointing along x (C) should have the same entropy as a collection pointing along z (A) rather than a random collection (B).

The correct answer is obtained if we first diagonalize $\psi\psi^\dagger$ and then use its diagonal elements (which are the eigenvalues) as the p_i's in Eq. (24.11). This is accomplished if we generalize Eq. (24.11) to write

$$\frac{S}{k} = -\text{Trace}\left[\rho \ln(\rho)\right] \tag{24.12}$$

where $\rho = \psi\psi^\dagger$ is a (2×2) matrix (called the density matrix).

24.5.2   *Does interaction increase the entropy?*

Back in Chapter 16 we discussed how a perfect anti-parallel (AP) spin valve could function like an info-battery (Fig. 16.3) that extracts energy from a collection of spins as it goes from the low entropy state A to the high entropy state B. But exactly how does this increase in entropy occur? In Chapter 16 we described the interaction as a "chemical reaction"

$$u + D \Leftrightarrow U + d \quad \text{(Same as Eq. (16.7))}$$

where u and d represent up and down channel electrons, while U and D represent up and down localized spins.

From a microscopic point of view the exchange interaction creates a superposition of wavefunctions as sketched on the next page.

We have shown equal superposition of the two possibilities for simplicity, but in general the coefficients could be any two complex numbers whose squared magnitude adds up to one.

Now the point is that the superposition state

$$\frac{1}{\sqrt{2}} u \times D + \frac{1}{\sqrt{2}} d \times U$$

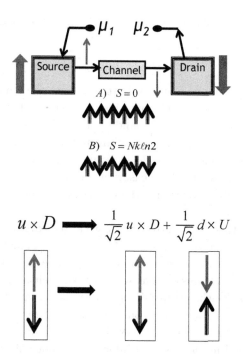

A) $S = 0$

B) $S = Nk\ell n2$

$$u \times D \longrightarrow \frac{1}{\sqrt{2}} u \times D + \frac{1}{\sqrt{2}} d \times U$$

has zero entropy just like the original state $u \times D$. Indeed we could picture a pseudo-spin whose up and down components are $u \times D$ and $d \times U$. The interaction merely rotates the pseudo-spin from the z to the x direction and as we discussed in the last section, mere rotation of spins or pseudo-spins causes no generation of entropy.

So how does the increase in entropy occur? The itinerant electron eventually gets extracted from the channel. At that moment there is a "collapse of the wavefunction" either into a $u \times D$ or a $d \times U$ depending on whether the channel electron is extracted by the source as an up electron or by the drain as a down electron. The localized spin is left behind in a down or an up state with 50% probability each. This is when the entropy increases by $k \ln(2)$.

24.5.3 *How much information can one spin carry?*

Suppose we decide to use the spin of the electron, that is the direction of the input magnet in Fig. 24.1 to convey information. It would seem that we could send large amounts of information, since there are now many possibilities. For example, suppose we choose a set of say 64 directions of the magnetization to convey information, it would seem that the entropy would be

$$S/k \ = \ \ln(64).$$

Note that we are using 64 figuratively to represent the number of magnetization directions we use, which could just as well be 10 or 100.

We have seen in Chapters 12 and 23 that a magnetic voltage probe making an angle θ with the injected spins measures a voltage proportional that depends on θ (Fig. 24.1) and it would seem that we could measure the direction of spin simply by measuring the voltage. This would allow us to encode 64 possible values of θ thereby transmitting $\ln(64)$ rather than $\ln(2)$.

But how can this be correct? Didn't we argue earlier that for one spin $S/k \ = \ \ln(2)$ rather than $\ln(64)$? These two arguments can be reconciled by noting that in order to measure a voltage that depends on θ we need many many electrons so that we can take their average. An individual electron would either transmit or not transmit into the magnet with specific probabilities that depend on θ. Only by averaging over many electrons would we get the average values that we have discussed. This means that we could send $\ln(64)$ worth of information, but only if we send many identically prepared electrons, so that the receiver can average over many measurements.

But couldn't we take one electron that we receive and create many electrons with the same wavefunction? After all, we can always copy a classical bit of information. There is a "no cloning theorem" that says we cannot copy the quantum state of an electron. The sender has to send us identically prepared electrons if we want to make many measurements and average.

These concepts are of course part of the field of quantum information on which much has been written and will be written. At the heart of this field is the *q-bit* and spin is its quintessential example.

Chapter 25

Epilogue:
Probabilistic Spin Logic (PSL)

Intel has a website, *From Sand to Circuits* (http://www.intel.com/about/companyinfo/museum/exhibits/sandtocircuits/index.htm) describing the amazing process that turns grains of sand into the chips that have enabled the modern information age. This book has been about the physics that these "grains of sand" and the associated technology have helped illuminate in the last 30 years, the physics that helped validate the concept of an elastic resistor with a clean separation of entropy-driven processes from the force-driven ones.

As we discussed in Chapter 1, the amazing progress of electronics has been enabled by continued downscaling whereby more transistors can be packed into a chip. The basic physics of field effect transistors (FET) has not changed much, though significant advances have been enabled by innovative geometry and/or materials. At this time devices incorporating new physics like tunneling FET's and negative capacitance FET's are also being explored to continue the downscaling.

Another field that has made great strides in the last two decades is that of spintronics which seeks to control the spin of electrons instead of traditional electronics based on the control of its charge. It has merged with what used to be a distinct field of research, namely nanomagnetics, to form a single thriving field which has had great impact on *memory devices*.

This book has been about electronic transport and in the last few chapters we have talked a little about spin transport. But a serious discussion of magnetics would take us too far afield. Nevertheless I would like to briefly introduce the reader to the remarkable synergistic relationship between magnets and spins, because they exemplify two key concepts in information theory, namely the *bit* which is at the heart of standard digital computing, and the *q-bit* which is at the heart of the emerging field of quantum computing. I would also like to use it to motivate a concept that is intermediate between a bit and a *q-bit*, which we call a *p-bit* (Camsari et al., 2017a).

25.1 Spins and Magnets

Loosely speaking, every electron is like an elementary magnet with a magnetic moment given by the Bohr magneton

$$\mu_B = \frac{q\hbar}{2m} = 9.27 \times 10^{-24} \text{A} \cdot \text{m}^2 \tag{25.1}$$

roughly what we would get if a current of 1 mA were circulating in a square loop with dimensions 0.1 nm × 0.1 nm loop. This was established back in the 1920s by the celebrated experiment due to Stern and Gerlach. More correctly the electron magnetic moment is given by

$$\mu_{el} = \frac{g_s}{2}\mu_B \tag{25.2}$$

g_s being the "g-factor" which is approximately equal to 2 for electrons in vacuum but could be significantly different in solids, just as the effective mass of electrons in solids can differ from that in vacuum. We will not worry about this "detail" and assume $g_s = 2$ for the following discussion.

If each electron is like a magnet then why are all materials not magnetic? Because usually the electrons are all paired with every up magnet balanced by a corresponding down magnet. It is only in magnetic materials like iron that internal interactions make it energetically favorable for a large number of electrons to line up in the same direction giving rise to a macroscopic magnetization whose magnitude is given by

$$M_s = \mu_B \frac{N_s}{\Omega} \tag{25.3}$$

N_s being the number of spins in a volume Ω. The magnitude of the magnetization of a magnet is fixed as long as the temperature T is well below its transition (Curie) temperature T_c. But its direction denoted by the unit vector $\hat{m}$ can change when a magnetic field is applied. The dynamics of $\hat{m}$ is described by the Landau-Lifshitz-Gilbert (LLG) equation which is widely used in the field of magnetics.

Which direction does the magnetization $\hat{m}$ point? Usually magnets have an easy axis, say the z-axis such that the magnet has two stable states $\pm\hat{z}$. If we put it along $+\hat{z}$ it will stay that way, and if we put it along $-\hat{z}$ it will stay that way.

But how do we put it along $+\hat{z}$ or $-\hat{z}$? The old method was to use a magnetic field in the z-direction. A more recent method is the use of *spin currents*: a large enough flux of z-directed electronic spins can also make the magnet switch just like a magnetic field (see Fig. 25.1a). This is one of the two key discoveries that have enabled the integration of nanomagnetics with spintronics that I mentioned. The other is the use of spin-valves and magnetic tunnel junctions (MTJ's) to READ (R) information for a magnet using the difference between the parallel (G_P) and anti-parallel (G_{AP}) conductances that we discussed earlier, see Fig. 23.6.

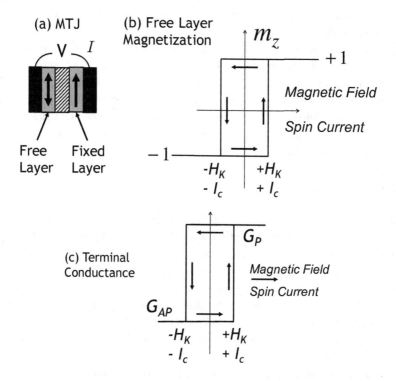

Fig. 25.1 (a) A magnet has an "easy axis" assumed to be along z. An external magnetic field or a spin current can be used to change its magnetization between -1 and $+1$ if it exceeds a critical field H_K. (b) A magnetic tunnel junction (MTJ) has one free layer whose magnetization can be flipped relatively easily compared to the fixed layer. (c) The change in the magnetization of the free layer translates into a change in the conductance G of the MTJ.

25.1.1   *Pseudospins and pseudomagnets*

Before moving on, let me mention an important lesson we can learn from the synergistic relationship between spins and magnets that has emerged in the last twenty years. We saw in Chapter 24 that diverse 2-component quantum objects can be viewed as two-component pseudo-spins (Section 24.4). Could we use these pseudospins to encode and transmit information? Perhaps, but our experience suggests that for information processing it is not enough to have a spin, we also need a magnet.

In standard charge-based architectures information is stored in capacitors and transmitted from capacitor to capacitor. Similarly we need a magnet to implement a spin capacitor and devices to transmit the information from magnet to magnet. Developments in the last decade have given us the basic ingredients. Whether we can build a information processing technology around it, remains to be seen.

It is tempting to go beyond simple spins and look at all kinds of exotic two-component pseudo-spins or even giant multi-component quasi-spins that maintain phase coherence over useful lengths of time. But it seems to me that a key question one should ask is, *"do we have a quasi-magnet to generate and detect the quasi-spin?"*

25.2   Unstable Magnets

Magnets with two stable states are routinely used to represent strings of 0's and 1's called bits in magnetic disks. A well-known problem with nano-magnets is that they can become unstable, unless the 0 and 1 states are separated by an energy barrier well in excess of the thermal energy kT. Indeed magnets have to be designed to have barriers $\sim 40\ kT$ in order to ensure that they can store information reliably for acceptable lengths of time, say a few years. If the barrier is $<< 40\ kT$ the magnet is unstable. At any given time they are either 0 or 1, but they continually fluctuate between the two.

It has been argued that this problem represents an opportunity. Low barrier magnets will switch back and forth in time between 0 and 1 so that their conductance fluctuates between G_P and G_{AP} (Fig. 25.1). Mathematically we can write

$$G_{MTJ} = G_{avg} + \Delta G \times \text{sgn}(\text{rand}(-1, +1)) \tag{25.4a}$$

$$\text{where } G_{avg} = \frac{G_P + G_{AP}}{2}, \quad \Delta G = \frac{G_P - G_{AP}}{2} \tag{25.4b}$$

and rand$(-1,+1)$ represents a random number between -1 and $+1$. The signum function (sign) converts negative values to -1 and positive values to $+1$. This describes the magnetization of a *telegraphic magnet* which fluctuates between -1 and $+1$, but does not take on intermediate values.

We can translate this conductance fluctuation into a voltage fluctuation using a potential divider circuit as shown in Fig. 25.2.

$$V = \frac{V_{DD}}{2} \ \frac{G_{MTJ} - G_L}{G_{MTJ} + G_L}. \tag{25.5a}$$

Since G_{MTJ} only takes on two values G_P and G_{AP}, we can write the voltage V in the form

$$V = V_{avg} + \Delta V \times \text{sgn}(\text{rand}(-1,+1)). \tag{25.5b}$$

This property has been used to build random binary number generators (RNG's) (see for example Grollier *et al.*, 2016).

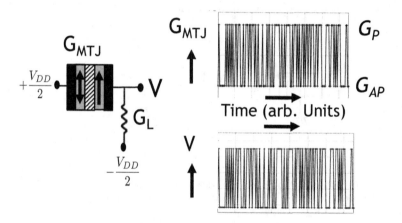

Fig. 25.2 Conductance of an MTJ with barrier height $<< 40 \ kT$ fluctuates with time and can be translated into a fluctuating voltage with a potential divider circuit as shown.

Stable magnets are routinely used to represent binary bits. At the other extreme, there is a large research activity in the area of quantum computers whose building blocks are q-bits represented by single spins that can exist in states that are neither '0' or '1', but rather a superposition of the two represented by a spinor $s = \{a \quad b\}^T$. Unstable magnets are either 0 or 1 at any given time, but they continually fluctuate between the two values, allowing us to implement something intermediate between a bit and a q-bit, what we call a probabilistic bit or a p-bit.

25.3 Three-terminal p-bits

The power of modern electronics lies not in the individual *bits* but in the amazing functionalities that are made possible by interconnecting large numbers of bits to build circuits. How can we interconnect p-bits to build p-circuits? We believe two-terminal p-bits like MTJ's are not suitable for the purpose. We need three-terminal transistor-like units, such that the stochastic output V_i of the ith p-bit is controlled by its input ν_i (see Fig. 25.3):

$$V_i(t + \Delta t) = \Delta V \times \text{sgn}\left(\text{rand}(-1, +1) + \tanh \frac{\nu_i(t)}{V_0}\right). \qquad (25.6a)$$

If the input voltage ν_i is zero, the output V_i fluctuates equally between -1 and $+1$ just like the 2-terminal p-bit. But the input voltage changes the relative probabilities, pinning it to $+1$ for large positive ν_i, and to -1 for large negative ν_i. The scale for what is large is set by the parameter V_0 which is assumed equal to $\Delta V/5$ in the plot.

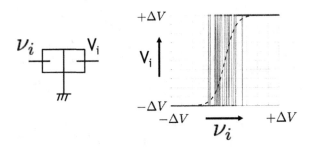

Fig. 25.3 Three-terminal p-bit or a p-transistor: An input voltage ν_i controls the stochastic output V_i of the ith p-bit as described mathematically by Eq. (25.6a). In this plot $V_0 = \Delta V/5$.

In short we now have a p-bit that *listens* to its input. If the input is derived from the outputs of other p-bits, they will become *correlated*. This can be achieved by interconnecting different p-bits with passive circuit elements like resistors or capacitors such that

$$\nu_i(t) = \sum_j J_{ij} V_j(t). \qquad (25.6b)$$

How would we build a three-terminal p-bit that follows Eq. (25.6a)? A relatively straightforward approach was described in Camsari (2017b) that combines a low-barrier MTJ with an ordinary transistor. Many other approaches should be possible, not necessarily based on magnets, and we will not go into the different possibilities. Instead let me conclude by addressing the motivational question: Why would we want to build p-circuits of the type described by Eqs. (25.6)?

25.4 p-circuits

Equations (25.6) are essentially the same as those used to describe stochastic neural networks which can be designed for all kinds of functions by choosing the weights J_{ij} appropriately. Rather than implement them with software simulations based on deterministic hardware, we could build p-circuits to *mimic* them in hardware which should be more efficient. Let me give an example.

A well-known concept in family trees is that of *relatedness* in family trees like the one shown in Fig. 25.4a. Consider two siblings F1 and F2 having the same parents GF and GM. Their relatedness is 50%, which we can understand by writing

$$F1 = 0.5 \ GF + 0.5 \ GM$$

$$F2 = 0.5 \ GF + 0.5 \ GM.$$

Assuming GF and GM are uncorrelated, we have

$$\langle F1 \times F2 \rangle = 0.25 \left(\langle GF \times GF \rangle + \langle GF \times GM \rangle + \langle GM \times GF \rangle + \langle GM \times GM \rangle \right)$$
$$= 0.25 \left(1 + 0 + 0 + 1 \right) = 0.5$$

where the angle brackets denote an average over many genes of GF and GM. If GF and GM are uncorrelated then they are as likely to have the same sign (product = +1) as to have the opposite sign (product = −1), so that the average value of $GF \times GM$ is zero. By contrast, both $GF \times GF$ and $GM \times GM$ are always +1.

Similarly the relatedness of first cousins C1 and C2 is known to be 12.5% which can be understood by writing

$$C1 = 0.5 \ F1 + 0.5 \ M1, \quad \text{and} \quad C2 = 0.5 \ F2 + 0.5 \ M2$$

so that

$$\langle C1 \times C2 \rangle = 0.25 \left(\langle F1 \times F2 \rangle + \langle F1 \times M2 \rangle + \langle M1 \times F2 \rangle + \langle M1 \times M2 \rangle \right)$$
$$= 0.25 \left(0.5 + 0 + 0 + 0 \right) = 0.125.$$

(a) Probabilistic network

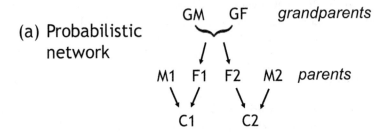

(b) *p*-circuit to mimic (a)

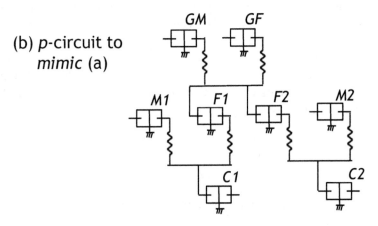

Fig. 25.4   (a) A well-known probabilistic network: A child C receives half its genetic information from each of the parents F,M who in turn get it from their parents. (b) A *p*-circuit composed of interconnected *p*-transistors designed to mimic the probabilistic network in (a).

Now the point is that a probabilistic network like this can be emulated with a circuit where each node is represented by a 3-terminal *p*-bit, and information is propagated from node i to node j by connecting the output V_i to the input v_j as shown in Fig. 25.4b. The *relatedness* of two physical nodes, A and B, can now be obtained by measuring the correlation between the voltages V at the corresponding circuit nodes

$$\langle V_A \times V_B \rangle = \int_0^T \frac{dt}{T} \, V_A(t) V_B(t).$$

The product of V_A and V_B can be obtained with a simple logic gate (specifically XNOR) and its output can be integrated with a long time constant ($\gg T$) RC circuit. Note that what we are measuring is a *time average* but it mimics the genetic relatedness which is an *ensemble average*.

I chose this example because of its familiar flavor, but probabilistic networks occur in many everyday problems, though they usually have 'children' with more than two 'parents', since a particular event is often influenced by more than two causes. In medical diagnosis, for example, multiple factors could lead to a particular symptom. To represent the causal relationships in general it may be necessary to introduce auxiliary nodes or 'hidden variables'. But let me not digress too far.

The point is that probabilistic nodes are common in everyday life and p-bits provide a natural representation. Moreover, probabilistic algorithms have proved very successful in machine learning, optimization and invertible logic among others. A p-computer should be able to implement such algorithms much more efficiently than a deterministic computer with artificially introduced randomness.

To build a p-computer we need three-terminal p-bits that can be interconnected into large scale p-circuits. Here we motivated the concept of p-bits with unstable magnets since spins and magnets span the full range from *bits* to *p-bits* to *q-bits*. But it is quite likely that as the field progresses, other physical realizations will emerge that go beyond spins and magnets, using different kinds of "pseudo-spins" and other generalized spin-like entities (Chapter 24) some of which could be based on charge.

In a seminal paper, *Simulating Physics with Computers* (Feynman, 1982) Feynman talked about probabilistic computers (p-computers) before going on to introduce the concept of quantum computers (q-computers), and highlighting the differences between the two. His vision arguably inspired the research into q-bits and q-computers. I would argue that p-bits and p-computers also deserve attention. They should operate robustly at room temperature with existing technology, unlike q-computers which require cryogenic temperatures to preserve quantum coherence. If p-computers can achieve even a fraction of the projected performance of q-computers, it should be at least a worthwhile intermediate step.

Perhaps, more intriguing is the possibility that a p-bit, could take digital computing to a new level by providing a natural building block for something that is more like the *human brain*.

Suggested Reading

This book is based on a set of two online courses originally offered in 2012 on nanoHUB-U and more recently in 2015 on edX. These courses are now available in self-paced format at nanoHUB-U (https://nanohub.org/u) along with many other unique online courses.

In preparing the second edition we decided to split the book into parts A and B following the two online courses available on nanoHUB-U entitled *Fundamentals of Nanoelectronics*

Part A: Basic Concepts   Part B: Quantum Transport.

Video lecture of possible interest in this context: NEGF: A Different Perspective. A detailed list of *video lectures* available at the course website corresponding to different sections of this volume (Part B: Quantum Transport) have been listed at the beginning.

Even this Second Edition represents lecture notes in unfinished form. I plan to keep posting additions/corrections at the book website.

This book is intended to be accessible to anyone in any branch of science or engineering, although we have discussed advanced concepts that should be of interest even to specialists, who are encouraged to look at my earlier books for additional technical details.

Datta S. (1995). Electronic Transport in Mesoscopic Systems
Datta S. (2005). Quantum Transport: Atom to Transistor
        Cambridge University Press

Over 50 years ago David Pines in his preface to the *Frontiers in Physics* lecture note series articulated the need for both a consistent account of a field and the presentation of a definite point of view concerning it. That is what we have tried to provide in this book, with no intent to slight any other point of view or perspective.

The viewpoint presented here is unique, but not the topics we discuss. Each topic has its own associated literature that we cannot do justice to. What follows is a *very incomplete list* representing a small subset of the relevant literature, consisting largely of references that came up in the text.

Chapter 1

Figure 1.5 is reproduced from

McLennan M. *et al.* (1991) Voltage Drop in Mesoscopic Systems, Phys. Rev. B, 43, 13846.

A recent example of experimental measurement of potential drop across nanoscale defects

Willke P. *et al.* (2015) Spatial Extent of a Landauer Residual-resistivity Dipole in Graphene Quantified by Scanning Tunnelling Potentiometry, Nature Communications, 6, 6399.

The transmission line model referenced in Section 1.8 is discussed in Section 9.4 and is based on

Salahuddin S. *et al.* (2005) Transport Effects on Signal Propagation in Quantum Wires, IEEE Trans. Electron Dev. 52, 1734.

Some of the classic references on the non-equilibrium Green's function (NEGF) method

Martin P.C. and Schwinger J. (1959) Theory of Many-particle Systems I, Phys. Rev. 115, 1342.

Kadanoff L.P. and Baym G. (1962) Quantum Statistical Mechanics, Frontiers in Physics, Lecture note series, Benjamin/Cummings.

Keldysh (1965) Diagram Technique for Non-equilibrium Processes, Sov. Phys. JETP 20, 1018.

The quote on the importance of the "channel" concept in Section 1.8 is taken from

Anderson P.W. (2010) 50 years of Anderson Localization, ed. E. Abrahams, Chapter 1, Thoughts on localization.

The quote on the importance of physical pictures, even if approximate, in Section 1.9 is taken from

Feynman R.P. (1963) Lectures on Physics, vol. II-2, Addison-Wesley.

Part 1: Contact-ing Schrödinger

Chapter 17

This discussion is based on Chapters 2–7 of Datta (2005).

For a detailed discussion of the self-consistent field method, the reader can look at

Herman F. and Skillman S. (1963) Atomic Structure Calculations, Prentice-Hall.
Slater J.C. (1963–1974) Quantum Theory of Molecules and Solids, Vols. I–IV, McGraw-Hill.

Chapter 18

This discussion is based on Chapter 8 of Datta (1995), and Chapters 8–10 of Datta (2005).

An experiment showing approximate conductance quantization in a hydrogen molecule.

Smit R.H.M. *et al.* (2002) Measurement of the Conductance of a Hydrogen Molecule, Nature 419, 906.

A standard text on NEGF

Haug H. and Jauho A.P. (1996) Quantum Kinetics in Transport and Optics of Semiconductors, Springer-Verlag.

The following reference is probably the first to apply NEGF to a device with contacts.

Caroli C. *et al.* (1972) A Direct Calculation of the Tunneling Current, J. Phys. C: Solid State Physics, 5, 21.

The NEGF method was related to the Landauer approach in

Datta, S. (1989) Steady-State Quantum Kinetic Equation, Phys. Rev., B40, 5830.

Meir Y. And Wingreen N. (1992) Landauer Formula for the Current through an Interacting Electron Region, Phys. Rev. Lett. 68, 2512.

Extensive numerical results based on the method of Datta (1989) are presented in

McLennan M. *et al.* (1991) Voltage Drop in Mesoscopic Systems, Phys. Rev. B, 43, 13846.

The dephasing model described in Section 18.4 is based on

R. Golizadeh-Mojarad *et al.* (2007), Non-equilibrium Green's function based model for dephasing in quantum transport, Phys. Rev. B 75, 081301 (2007).

A glimpse of the impressive progress in NEGF-based device modeling

Steiger S. *et al.* (2011) NEMO5: A Parallel Multiscale Nanoelectronics Modeling Tool, IEEE Transactions on Nanotechnology, 10, 1464.

Chapter 19

A paper on localization by the person who pioneered the field along with many other seminal concepts.

Anderson P.W. *et al.* (1981) New Method for a Scaling Theory of Localization, Phys. Rev. B 23, 4828.

Resonant tunneling is discussed in more detail in Chapter 6 of Datta (1995) and Chapter 9 of Datta (2005).

Part 2: More on NEGF

Chapter 20

Two experiments reporting the discovery of quantized conductance in ballistic conductors.

van Wees, B.J. *et al.* (1988) Quantized Conductance of Points Contacts in a Two-Dimensional Electron Gas, Phys. Rev. Lett. 60, 848.

Wharam, D.A. *et al.* (1988) One-Dimensional Transport and the Quantisation of the Ballistic Resistance, J. Phys. C. 21, L209.

An instructive paper on graphene

Brey, L., and H. Fertig, (2006) Edge States and Quantized Hall Effect in Graphene, Phys. Rev. B 73, 195408.

A recent book with a thorough discussion of graphene-related materials

Torres L.E.F. Foa *et al.* (2014) Introduction to Graphene-based Nanomaterials, Cambridge University Press.

The paper that reported the first observation of the amazing quantization of the Hall resistance:

von Klitzing K. *et al.* (1980) New Method for High-Accuracy Determination of the Fine Structure Constant Based on Quantized Hall Resistance, Phys. Rev. Lett. 45, 494.

For more on edge states in the quantum Hall regime the reader could look at Chapter 4 of Datta (1995) and references therein.

Chapter 21

A couple of classic references on the NEGF treatment of interactions

Danielewicz, P. (1984) Quantum Theory of Non-Equilibrium Processes, Ann. Phys., NY, 152, 239.

Mahan G.D. (1987) Quantum Transport Equation for Electric and Magnetic Fields, Phys. Rep. NY, 145, 251.

Readers interested in device analysis at high bias may find this article useful. MATLAB codes available on our website.

Datta S. (2000) Nanoscale Device Modeling: The Green's Function Method, Superlattices and Microstructures, 28, 253.

Chapter 22

The rate equations described here are based on

Beenakker C.W.J. (1991) Theory of Coulomb Blockade Oscillations in the Conductance of a Quantum Dot, Phys. Rev. B, 44, 1646.

The reader may also find this reference helpful

Bonet E., Deshmukh M.M., and Ralph D.C. (2002) Solving Rate Equations for Electron Tunneling via Discrete Quantum States, Phys. Rev. B 65, 045317.

For an application of the rate equations to describe interesting current-voltage characteristics observed in double quantum dots

Muralidharan B. and Datta S. (2007) Generic model for current collapse in spin-blockaded transport Phys. Rev. B76, 035432.

For a description of methods that go beyond the simple rate equation, the reader could look at

M. Braun, J. Koenig, and J. Martinek (2004) Theory of Transport through Quantum-Dot Spin Valves in the Weak Coupling Regime, Phys. Rev. B, 70, 195345.

Braig S. and Brouwer P.W. (2005) Rate Equations for Coulomb Blockade with Ferromagnetic Leads, Phys. Rev. B 71, 195324 (2005).

Part 3: Spin Transport

Chapter 23

Most quantum mechanics texts have a thorough discussion of spinors. A non-standard reference that the reader may enjoy

Misner C.W., Thorne K.S. and Wheeler J.A. (1970) Gravitation, Chapter 41, Freeman.

A textbook written from an applied perspective

Bandyopadhyay, S. and Cahay, M. (2008) Introduction to Spintronics, Taylor & Francis.

An article on spin injection by one of the inventors of the spin valve:

Fert A. *et al.* (2007) Semiconductors between Spin-Polarized Sources and Drains, IEEE Trans. Electron Devices 54, 921.

A review article

Schmidt G. (2005) Concepts for Spin Injection into Semiconductors a Review J. Phys. D: Appl. Phys. 38, R107.

One of the widely cited papers on non-local spin voltages

Takahashi S. and Maekawa S. (2003) Spin Injection and Detection in Magnetic Nanostructures, Phys. Rev. B 67, 052409.

A few interesting experiments showing the effects of spin coherence

Huang B. *et al.* (2007) Geometric Dephasing-limited Hanle Effect in Long Distance Lateral Silicon Spin Transport Devices, Appl. Phys. Lett. 93, 162508.

Sih V. *et al.* (2006) Generating Spin Currents in Semiconductors with the Spin Hall Effect, Phys. Rev. Lett. 97, 2096605.

A few papers using the NEGF method to model spin transport.

Nikolic B. *et al.* (2010) Spin Currents in Semiconductor Nanostructures: A Non-Equilibrium Green-Function Approach, Chapter 24, The Oxford Handbook on Nanoscience and Technology: Frontiers and Advances, eds. A. V. Narlikar and Y. Y. Fu, Oxford University Press (available on condmat, arXiv:0907.4122). See also Chapter 3 by R. Golizadeh-Mojarad *et al.* and Datta (2010).

Datta D. *et al.* (2012) Voltage Asymmetry of Spin Transfer Torques, IEEE Transactions on Nanotechnology, 11, 261.

Zainuddin A.N.M. *et al.* (2011) Voltage-controlled Spin Precession, Phys. Rev. B 84, 165306.

An experimental paper showing evidence for spin-polarized surface states, including references to the literature on topological insulators.

Xiu F. *et al.* (2011) Manipulating Surface States in Topological Insulator Nanoribbons, Nature Nanotechnology 6, 216 and references therein.

Electrical control of spins, or the so-called Datta-Das oscillations

Datta S. and Das B. (1990) Electronic analog of the electro-optic modulator, Appl. Phys. Lett., 56, 665.
Koo H.C. *et al.* (2009) Control of Spin Precession in a Spin-Injected Field Effect Transistor, Science 325, 1515.
Wunderlich *et al.* (2010) Spin Hall Effect Transistor, Science 330, 1801.
Choi *et al.* Electrical detection of coherent spin precession using the ballistic intrinsic spin Hall effect (2015) Nat. Nanotech. 10, 666.

A classic reference describing the BdG equation.

P.G. de Gennes (1968) Superconductivity of Metals and Alloys, Advanced Book Classics (1968).

A couple of references that could help the reader apply the NEGF discussed here to problems involving Andreev scattering using the (2×2) BdG equation.

Datta S. and Bagwell P.F. (1999) Can the Bogoliubov-deGennes equation be interpreted as a one-particle wave equation? (1999) *Superlattices and Microstructures*, 25, 1233.
Samanta M.P. and Datta S. (1998) Electrical transport in junctions between unconventional superconductors: Application of the Green's-function formalism *Phys. Rev. B* 57, 10972 (1997).

A couple of references that use the (4×4) BdG equation, which also provide a lead to the vast literature on Majorana modes.

Pikulin D.I. *et al.* (2012) A zero-voltage conductance peak from weak antilocalization in a Majorana nanowire, *New Journal of Physics* 14, 125011.

P. San-Jose (2013) Multiple Andreev reflection and critical current in topological superconducting nanowire junctions. *New Journal of Physics* 15, 075019 (2013).

Chapter 24

A spin diffusion equation based on the Kubo formalism is described in

Burkov *et al.* (2004) Theory of spin-charge-coupled transport in a two-dimensional electron gas with Rashba spin-orbit interactions, Phys. Rev. B 70, 155308.

The approach described here is based on

Sayed *et al.* (2017) Transmission Line Model for Charge and Spin Transport in Channels with Spin-Momentum Locking, https://arxiv.org/abs/1707.04051.

Spin circuits provide an attractive approximation to the full NEGF model that may be sufficiently accurate in many cases. See for example

Brataas A. *et al.* (2006) Non-Collinear Magnetoelectronics, Phys. Rep., 427, 157.

Srinivasan S. *et al.* (2011) All-Spin Logic Device With Inbuilt Nonreciprocity, IEEE Transactions On Magnetics, 47, 4026.

A paper stressing the distinction between transport spin currents and equilibrium spin currents.

Rashba E.I. (2003) Spin Currents in Thermodynamic Equilibrium: The Challenge of Discerning Transport Currents, Phys. Rev. B 68, 241315R.

A well-cited paper that discusses some issues of quantum information we touched on

Zurek W. (2003) Decoherence, Einselection and the Quantum Origins of the Classical, Rev. Mod. Phys., 75, 715.

Chapter 25

A couple of lucid references that could help introduce the reader to the LLG Equation and its use in describing the dynamics of magnets.

Sun J.Z. (2000) Spin-current interaction with a monodomain magnetic body: A model study, Phys. Rev. B, 62, 570, see references.

Butler W. H. *et al.* (2012) Switching distributions for perpendicular spin-torque devices within the macrospin approximation. IEEE Trans. Magn. 48, 4684.

The negative capacitance transistor is based on ferroelectrics, which are similar to ferromagnets in some ways, but it is based on charge rather than spin. An old reference and a very recent one

Salahuddin S. and Datta S. (2008) Use of Negative Capacitance to Provide Voltage Amplification for Ultralow Power Nanoscale Devices, Nanoletters, 8, 405.

Chatterjee K., Rosner A.J. and Salahuddin S. (2017) Intrinsic speed limit of negative capacitance transistors, IEEE Trans. Electron Device Letters, 38, 1328.

Our Eqs. (25.6) describe a subset of the area of stochastic neuromorphic computing. For a general survey of neuromorphic computing, the reader could look at

Schuman C.D. *et al.* A survey of neuromorphic computing and neural networks in hardware, https://arxiv.org/abs/1705.06963.

A couple of papers highlighting the stochastic behavior of nanomagnets and their possible applications.

Locatelli N. *et al.* (2014) Noise-Enhanced Synchronization of Stochastic Magnetic Oscillators, Phys. Rev. Applied 2, 034009.

Grollier J. *et al.* (2016) Spintronic Nanodevices for Bioinspired Computing, Proc. IEEE 104, 2024.

Here are three of our papers proposing different applications of PSL using p-bits:

Ising computing and Bayesian inference: Behin-Aein B., Diep V. and Datta S. (2016), A Building Block for Hardware Belief Networks, Scientific Reports 6, 29893.

Traveling salesman problem: Sutton B.M., Camsari K.Y., Behin-Aein B. and Datta S. (2017) Intrinsic optimization using stochastic nano-magnets, Scientific Reports, 7, 44370.

Invertible Boolean logic: Camsari K.Y., Faria R., Sutton B.M. and Datta S. (2017a) Stochastic p-bits for Invertible Boolean Logic, Phys. Rev. X, 3, 031014.

This paper proposes a way to implement a three-terminal p-bit by combining an MTJ with a standard transistor.

Camsari K.Y., Salahuddin S. and Datta S. (2017b) Implementing p-bits with Embedded MTJ, IEEE Trans. Electron Device Letters, 38, 1767.

In this seminal paper which inspired the field of quantum computing, Feynman talks about probabilistic computers before introducing quantum computers.

Feynman R.P. (1982) Simulating physics with computers, International Journal of Theoretical Physics, 21, 467.

PART 4
Appendices

Appendix F

List of Equations and Figures Cited From Part A

The equations cited here from Part A (Basic Concepts) of this book are listed in this appendix for the convenience of the readers. If the meaning of a symbol is not clear from the context, the list of symbols at the beginning of this book may be useful.

Cited Equations

Eq. (2.11): $\quad f(E) - f_0(E) \approx \left(-\dfrac{\partial f_0}{\partial E}\right)(\mu - \mu_0)$

Eq. (3.1): $\quad \dfrac{I}{V} = \displaystyle\int_{-\infty}^{+\infty} dE \left(-\dfrac{\partial f_0}{\partial E}\right) G(E)$

Eq. (3.3): $\quad I = \dfrac{1}{q} \displaystyle\int_{-\infty}^{+\infty} dE\, G(E)\, (f_1(E) - f_2(E))$

Eq. (6.17): $\quad N(p) = \left\{ 2\dfrac{L}{L/p}, \quad \pi\dfrac{LW}{(L/p)^2}, \quad \dfrac{4\pi}{3}\dfrac{LA}{(L/p)^3} \right\}$

Eq. (9.5): $\quad \dfrac{\partial f}{\partial t} + v_z\dfrac{\partial f}{\partial z} + F_z\dfrac{\partial f}{\partial p_z} = S_{op}f$

Eq. (10.1): $\quad G_{4t} = \dfrac{I}{(\mu_1^* - \mu_2^*)/q} = M\dfrac{q^2}{h}\dfrac{T}{1-T}$

Eq. (10.2): $\quad G_{2t} = \dfrac{I}{(\mu_1 - \mu_2)/q} = M\dfrac{q^2}{h}T$

Eq. (10.3): $\quad I_m = \dfrac{1}{q}\displaystyle\sum_m G_{m,n}\,(\mu_m - \mu_n)$

Eq. (11.5): $\quad \dfrac{V_H}{W} = v_d B$

Eq. (11.6): $\qquad R_H = \dfrac{V_H}{I} = \dfrac{B}{q(N/LW)}$

Eq. (11.10): $\qquad \omega_c = \left|\dfrac{qvB}{p}\right|_{E=\mu_0} = \left|\dfrac{qB}{m}\right|_{E=\mu_0}$

Eq. (12.18a): $\qquad \mu = \dfrac{\mu^{up} + \mu^{dn}}{2}$

Eq. (12.18b): $\qquad \mu_S = \dfrac{\mu^{up} - \mu^{dn}}{2}$

Eq. (12.19a): $\qquad \mu_P = \mu + \dfrac{P\mu_S}{2}$

Eq. (12.19b): $\qquad P = \dfrac{g^{up} - g^{dn}}{g^{up} + g^{dn}}$

Eq. (12.24): $\qquad I = G_B \dfrac{\mu^+ - \mu^-}{q}$

Eq. (15.18): $\qquad p_i = \dfrac{1}{Z} e^{-(E_i - \mu N_i)/kT}$

Eq. (16.7): $\qquad u + D \Longleftrightarrow U + d$

Cited Figures

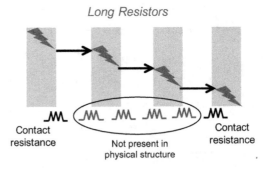

Long Resistors

Contact resistance   Not present in physical structure   Contact resistance

Fig. F.1 (Same as Fig. 3.5, Part A) A long resistor can be viewed as a series of ideal elastic resistors. However, we have to exclude the resistance due to all the conceptual interfaces that we introduce which are not present in the physical structure.

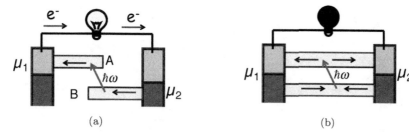

Fig. F.2 (Same as Fig. 12.6, Part A) (a) Asymmetric contacts are central to the operation of the "solar cell". (b) If contacted symmetrically no electrical output is obtained.

Appendix G

NEGF Equations

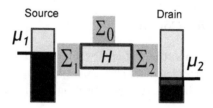

"Input":

(1) **H**-matrix parameters chosen appropriately to match energy levels or dispersion relations.
(2) Procedure for obtaining $\boldsymbol{\Sigma}_m$ for terminal m is summarized below in Section G.1.

$$\boldsymbol{\Sigma} = \boldsymbol{\Sigma}_0 + \boldsymbol{\Sigma}_1 + \boldsymbol{\Sigma}_2 + \cdots$$

$$\boldsymbol{\Sigma}^{in} = \boldsymbol{\Sigma}_0^{in} + \boldsymbol{\Sigma}_1^{in} + \boldsymbol{\Sigma}_2^{in} + \cdots$$

$$\boldsymbol{\Gamma}_j = i\left[\boldsymbol{\Sigma}_j - \boldsymbol{\Sigma}_j^\dagger\right], \quad j = 0, 1, 2, \cdots$$

"Required Equations":

1. Green's Function:

$$\mathbf{G}^R = [E\mathbf{I} - \mathbf{H} - \boldsymbol{\Sigma}]^{-1} \quad \text{and} \quad \mathbf{G}^A = \left[\mathbf{G}^R\right]^\dagger \quad \text{(same as Eq. (18.1))}$$

2. "Electron Density" times 2π:

$$\mathbf{G}^n = \mathbf{G}^R \boldsymbol{\Sigma}^{in} \mathbf{G}^A \quad \text{(same as Eq. (18.2))}$$

3. "Density of states" times 2π:

$$\mathbf{A} = \mathbf{G}^R \mathbf{\Gamma} \, \mathbf{G}^A = \mathbf{G}^A \mathbf{\Gamma} \, \mathbf{G}^R = i[\mathbf{G}^R - \mathbf{G}^A] \quad \text{(same as Eq. (18.3a))}$$

4a. Current / energy at terminal "m":

$$\tilde{I}_m = \frac{q}{h} \, \text{Trace} \, [\mathbf{\Sigma}_m^{in} \mathbf{A} - \mathbf{\Gamma}_m \, \mathbf{G}^n] \quad \text{(same as Eq. (18.4))}$$

4b. Current / energy at terminal "m", to be used only if Σ_0 is zero

$$I_m = \frac{q}{h} \sum_n \bar{T}_{mn} \, (f_m(E) - f_n(E)) \quad \text{(same as Eq. (18.31))}$$

$$\bar{T}_{mn} \equiv \quad \text{Trace} \, [\mathbf{\Gamma}_m \mathbf{G}^R \mathbf{\Gamma}_n \mathbf{G}^A] \quad \text{(same as Eq. (18.32))}$$

G.1   Self-energy for Contacts

(1) For 1D problems, the self-energy function for each contact has a single non-zero element $t \, e^{ika}$ corresponding to the point that is connected to that contact (see Section 19.1).

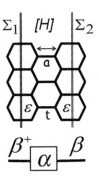

(2) 2D Hamiltonians for any conductor with a uniform cross-section can be visualized as a linear 1D chain of "atoms" each having an on-site matrix Hamiltonian $\boldsymbol{\alpha}$ coupled to the next "atom" by a matrix $\boldsymbol{\beta}$.

Each of the matrices $\boldsymbol{\alpha}$ and $\boldsymbol{\beta}$ is of size $(n \times n)$, n being the number of basis functions describing each unit.

The ***self-energy matrix*** Σ_m for terminal m is zero except for the last $(n \times n)$ block at the surface. This block is obtained from

$$\boldsymbol{\beta} \, \mathbf{g} \, \boldsymbol{\beta}^\dagger \quad \text{(Same as Eq. (20.6a))}$$

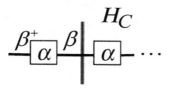

where the surface Green function **g** is calculated from a recursive relation:

$$\mathbf{g}^{-1} = (E + i0^+)\,\mathbf{I} - \boldsymbol{\alpha} - \boldsymbol{\beta}\,\mathbf{g}\,\boldsymbol{\beta}^\dagger \qquad \text{(Same as Eq. (20.6b))}$$

In the rest of this Appendix, we will obtain Eqs. (20.6). To obtain these results, first we consider just the last point of the device and its connection to the infinite contact described by $\mathbf{H}_c$:

$$\begin{bmatrix} \boldsymbol{\alpha} & \mathbf{B} \\ \mathbf{B}^\dagger & \mathbf{H}_c \end{bmatrix}$$

where $\mathbf{B} \equiv \begin{bmatrix} \boldsymbol{\beta} & 0 & 0 & \cdots \end{bmatrix}$.

The overall Green's function can be written as

$$\begin{bmatrix} \mathbf{A} & -\mathbf{B} \\ -\mathbf{B}^\dagger & \mathbf{A}_c \end{bmatrix}^{-1} \equiv \begin{bmatrix} \mathbf{G}^R & \cdots \\ \cdots & \cdots \end{bmatrix}$$

where

$$\mathbf{A} \equiv (E + i0^+)\mathbf{I} - \boldsymbol{\alpha} \tag{G.1}$$

$$\mathbf{A}_c \equiv (E + i0^+)\mathbf{I}_c - \mathbf{H}_c \tag{G.2}$$

With a little matrix algebra we can show that the top block of the Green's function, $\mathbf{G}^R$ is given by

$$\mathbf{G}^R = [\mathbf{A} - \mathbf{B}\,\mathbf{A}_c^{-1}\,\mathbf{B}^\dagger]^{-1} \tag{G.3}$$

so that we can identify self-energy as

$$\boldsymbol{\Sigma} = \mathbf{B}\mathbf{A}_c^{-1}\mathbf{B}^\dagger.$$

Since $\mathbf{B}$ has only one non-zero element $\boldsymbol{\beta}$, we can write

$$\boldsymbol{\Sigma} = \boldsymbol{\beta}\mathbf{g}\boldsymbol{\beta}^\dagger \qquad \text{(same as Eq. (20.6a))}$$

where **g** represents the top block of $[\mathbf{A}_c]^{-1}$, often

$$\begin{bmatrix} \mathbf{A} & -\boldsymbol{\beta} & 0 & 0 & \cdots \\ -\boldsymbol{\beta}^\dagger & \mathbf{A} & -\boldsymbol{\beta} & 0 & \cdots \\ 0 & -\boldsymbol{\beta}^\dagger & \mathbf{A} & -\boldsymbol{\beta} & \cdots \\ \vdots & \vdots & \vdots & \vdots & \ddots \end{bmatrix}^{-1} \equiv \begin{bmatrix} \mathbf{g} & \cdots & \cdots & \cdots & \cdots \\ \cdots & \cdots & \cdots & \cdots & \cdots \\ \cdots & \cdots & \cdots & \cdots & \cdots \\ \vdots & \vdots & \vdots & \vdots & \ddots \end{bmatrix} \tag{G.4}$$

To obtain Eq. (20.6b), we apply Eq. (G.3) to the $(N \times N)$ matrix in Eq. (G.4) treating the first block $\mathbf{A}$ as the "device", and the rest of the $(N-1) \times (N-1)$ as contact, to obtain

$$\mathbf{g}_N = [\mathbf{A} - \boldsymbol{\beta}\mathbf{g}_{N-1}\boldsymbol{\beta}^\dagger]^{-1} \tag{G.5}$$

where $\mathbf{g}_N$ represents the $\mathbf{g}$ on the right-hand side of Eq. (G.4) if the matrix on the left is of size $N \times N$. One could solve Eq. (G.5) recursively starting from $\mathbf{g}_1$ to $\mathbf{g}_2$ and so on till $\mathbf{g}_N$ is essentially the same as $\mathbf{g}_{N-1}$. At that point we have the solution to Eq. (20.6b)

$$\mathbf{g} = [\mathbf{A} - \boldsymbol{\beta}\,\mathbf{g}\,\boldsymbol{\beta}^\dagger]^{-1} \quad \text{(Similar to Eq. (20.6b))}$$

G.2   Self-energy for Elastic Scatterers in Equilibrium:

$$\Sigma_{kl} = D_{kl,\,ij}\, G_{ij}, \qquad [\Sigma^{in}]_{kl} = D_{kl,\,ij}\, [G^n]_{ij} \tag{G.6}$$

where summation over repeated indices is implied and

$$D_{kl,\,ij} = \left\langle [U_s]_{ki}\, [U_s]^*_{lj} \right\rangle \tag{G.7}$$

where $\langle \cdots \rangle$ denotes average value, and we are considering a general scattering potential with non-zero off-diagonal elements.

In Chapter 18 we assumed that only the diagonal elements are non-zero then the $\mathbf{D}$ can be simplified from a fourth-order tensor to a second-order tensor or in other words a matrix

$$D_{kl} = \left\langle [U_s]_{kk}\, [U_s]^*_{ll} \right\rangle. \tag{G.8}$$

In terms of this matrix, Eq. (G.6) can be rewritten as

$$\Sigma_{kl} = D_{kl}\, G_{kl}, \qquad [\Sigma^{in}]_{kl} = D_{kl}\, [G^n]_{kl} \tag{G.9}$$

which amounts to an element by element multiplication:

$$\boldsymbol{\Sigma} = \mathbf{D} \times \mathbf{G}, \quad \boldsymbol{\Sigma}^{in} = \mathbf{D} \times \mathbf{G}^n \quad \text{(see Eqs. (18.35))}$$

G.3   Self-energy for Inelastic Scatterers

(*See Chapter 21, and also Datta(2005)*)

$$[\Sigma^{in}]_{kl} = D_{kl,\,ij}\,(\hbar\omega)\, [G^n(E - \hbar\omega)]_{ij} \tag{G.10a}$$

$$\Gamma_{kl} = D_{kl,\,ij}\,(\hbar\omega)\, [G^n(E - \hbar\omega) + G^p(E + \hbar\omega)]_{ij} \tag{G.10b}$$

where $\mathbf{G}^p = \mathbf{A} - \mathbf{G}^n$, and summation over repeated indices is implied.

$$\Sigma_{kl} = \underbrace{h_{kl}}_{\substack{\text{Hilbert} \\ \text{Transform} \\ \text{of } \Gamma_{kl}}} - \frac{i}{2} \Gamma_{kl} \qquad \text{(G.10c)}$$

For scatterers in equilibrium with temperature T_s,

$$\frac{D_{kl,ij}(+\hbar\omega)}{D_{ji,lk}(-\hbar\omega)} = e^{-\hbar\omega/kT_s} \qquad \text{(G.11)}$$

Appendix H

MATLAB Codes Used for Text Figures

These codes are included here mainly for their pedagogical value. It is planned to make soft copies available through our website Link of codes.

H.1 Chapter 19

H.1.1 *Fig. 19.2 Transmission through a single point scatterer in a 1D wire*

```
clear all

t0=1;Np=11;X=[0:1:Np-1];
L=diag([1 zeros(1,Np-1)]);R=diag([zeros(1,Np-1) 1])
    ;
zplus=i*1e-12;

H0=2*t0*diag(ones(1,Np))-t0*diag(ones(1,Np-1),1)...
-t0*diag(ones(1,Np-1),-1);
N1=3;N2=9;UB1=2*t0;UB2=0*2*t0;
H0(N1,N1)=H0(N1,N1)+UB1;H0(N2,N2)=H0(N2,N2)+UB2;H=
    H0;

VV=0;UV=linspace(0,-VV,Np);%Linear Potential

ii=1;dE=5e-2; for EE=[-.5:dE:4.5]*t0
    ck=(1-(EE-UV(1)+zplus)/(2*t0));ka=acos(ck);
s1=-t0*exp(i*ka);sig1=kron(L,s1);
```

199

```
17        ck=(1-(EE-UV(Np)+zplus)/(2*t0)));ka=acos(ck)
            ;
18        s2=-t0*exp(i*ka);sig2=kron(R,s2);
19     gam1=i*(sig1-sig1');gam2=i*(sig2-sig2');
20
21     G=inv((EE*eye(Np))-H-diag(UV)-sig1-sig2);
22     Tcoh(ii)=real(trace(gam1*G*gam2*G'));
23     E(ii)=EE/t0;ii=ii+1;
24  end
25
26  hold on
27  h=plot(Tcoh,E,'k-o');
28  set(h,'linewidth',[1.2])
29  set(gca,'Fontsize',[36])
30  axis([-0.1 1.1 -.5 4.5]);
31  grid on
```

H.1.2   *Fig. 19.4 Normalized conductance for a wire with*
$M = 1$ due to one scatterer

```
1  clear all
2
3  t0=1;Np=11;X=[0:1:Np-1];
4  L=diag([1 zeros(1,Np-1)]);R=diag([zeros(1,Np-1) 1])
      ;
5  zplus=i*1e-12;
6
7  H0=2*t0*diag(ones(1,Np))-t0*diag(ones(1,Np-1),1)...
8  -t0*diag(ones(1,Np-1),-1);
9  N1=3;N2=9;UB1=2*t0;UB2=2*t0;
10 H0(N1,N1)=H0(N1,N1)+UB1;H0(N2,N2)=H0(N2,N2)+UB2;H=
      H0;
11
12 VV=0;UV=linspace(0,-VV,Np); % Linear potential
13
14 ii=1;dE=5e-4; for EE=[-.25:dE:1.25]*t0
15     ck=(1-(EE-UV(1)+zplus)/(2*t0));ka=acos(ck);
16     theta(ii)=(real(ka)*(N2-N1+1)/pi);
```

```
17    s1=-t0*exp(i*ka);sig1=kron(L,s1);
18        ck=(1-(EE-UV(Np)+zplus)/(2*t0));ka=acos(ck)
              ;
19        s2=-t0*exp(i*ka);sig2=kron(R,s2);
20    gam1=i*(sig1-sig1');gam2=i*(sig2-sig2');
21
22    G=inv((EE*eye(Np))-H-diag(UV)-sig1-sig2);
23    Tcoh(ii)=real(trace(gam1*G*gam2*G'));
24    E(ii)=EE/t0;ii=ii+1;
25 end
26
27 hold on
28 h=plot(Tcoh,E,'k');
29 set(h,'linewidth',[3.0])
30 set(gca,'Fontsize',[36])
31 axis([-0.1 1.1 -.25 1.25]);
32 grid on
```

H.1.3 *Fig. 19.5 Normalized conductance for a wire with $M = 1$ due to six scatterers*

```
1  clear all
2  t0=1;
3  Np=100;Np1=11;Np2=23;Np3=7;Np4=31;
4  L=diag([1 zeros(1,Np-1)]);R=diag([zeros(1,Np-1) 1])
      ;
5  zplus=i*1e-12;
6
7  H0=2*t0*diag(ones(1,Np))-t0*diag(ones(1,Np-1),1)...
8  -t0*diag(ones(1,Np-1),-1);
9  UB=2*t0;n=1;
10 H=H0+UB*diag([n zeros(1,Np1) 1 zeros(1,Np2) n zeros
      (1,Np3) n ...
11     zeros(1,Np4) n zeros(1,Np-Np1-Np2-Np3-Np4-6) n
          ]);
12
13 ii=1; for EE=[-.25:1e-3:1]*t0
14 % for EE=t0:-dE:t0
```

```
15    ck=(1-(EE+zplus)/(2*t0));ka=acos(ck);
16    s1=-t0*exp(i*ka);s2=-t0*exp(i*ka);
17    sig1=kron(L,s1);sig2=kron(R,s2);
18    gam1=i*(sig1-sig1');gam2=i*(sig2-sig2');
19
20    G=inv((EE*eye(Np))-H-sig1-sig2);
21    A=real(diag(i*(G-G')));ii
22       Gn=G*gam1*G';
23
24    Tcoh(ii)=real(trace(gam1*G*gam2*G'));TM(ii)=
          real(trace(gam2*Gn));
25    E(ii)=EE/t0;ii=ii+1;
26  end
27
28  hold on
29  %h=plot(Tcoh./(6.-5*Tcoh),E,'k-o');
30  h=plot(Tcoh,E,'k');
31  set(h,'linewidth',[3.0])
32  set(gca,'Fontsize',[36])
33  axis([-.1 1.1 -.25 1])
34  grid on
```

H.1.4   *Figs. 19.6–19.7 Potential drop across a scatterer
            calculated from NEGF*

```
1   clear all
2
3   t0=1;Np=51;X=[0:1:Np-1];Nh=floor(Np/2);
4   L=diag([1 zeros(1,Np-1)]);R=diag([zeros(1,Np-1) 1])
      ;
5   zplus=i*1e-12;D=9e-2*t0^2;
6
7   sigB=zeros(Np);siginB=zeros(Np);
8
9   H0=2*t0*diag(ones(1,Np))-t0*diag(ones(1,Np-1),1)...
10  -t0*diag(ones(1,Np-1),-1);
11  N1=Nh+1;UB1=1*t0;
12  H0(N1,N1)=H0(N1,N1)+UB1;H=H0;
```

```
13
14  EE=t0;
15  ck=(1-(EE+zplus)/(2*t0));ka=acos(ck);
16  v=2*t0*sin(ka);
17  % Semiclassical profile
18  T=real(v^2/(UB1^2+v^2));R1=(1-T)/T;
19  TT=real(v^2/(D+v^2));R2=1*(1-TT)/TT;
20  RR=[0.5 R2*ones(1,Nh) R1 R2*ones(1,Nh) 0.5];
21  RR=cumsum(RR);Vx=ones(1,Np+2)-(RR./RR(Np+2));
22  Fclass=Vx([2:Np+1]);
23  %Based on resistance estimates
24
25  s1=-t0*exp(i*ka);sig1=kron(L,s1);
26  ck=(1-(EE+zplus)/(2*t0));ka=acos(ck);
27  s2=-t0*exp(i*ka);sig2=kron(R,s2);
28  gam1=i*(sig1-sig1');gam2=i*(sig2-sig2');
29
30  G=inv((EE*eye(Np))-H-sig1-sig2);
31  Tcoh=real(trace(gam1*G*gam2*G'));
32
33  change=100;
34  while change>1e-6
35  G=inv((EE*eye(Np))-H-sig1-sig2-sigB);
36  sigBnew=diag(diag(D*G));%sigBnew=D*G;
37  change=sum(sum(abs(sigBnew-sigB)));
38  sigB=sigB+0.25*(sigBnew-sigB);
39  end
40  A=real(diag(i*(G-G')));change=100;
41  while change>1e-6
42  Gn=G*(gam1+siginB)*G';
43  siginBnew=diag(diag(D*Gn));%siginBnew=D*Gn;
44  change=sum(sum(abs(siginBnew-siginB)));
45  siginB=siginB+0.25*(siginBnew-siginB);
46  end
47  F=real(diag(Gn))./A;
48
49  hold on
50  h=plot(X,F,'k');
```

```
51  set(h,'linewidth',[3.0])
52  h=plot(X,Fclass,'r-o');
53  set(h,'linewidth',[1.2])
54  set(gca,'Fontsize',[36])
55  xlabel(' z ---> ')
56  grid on
57  axis([-10 60 0 1])
```

H.1.5 *Figs. 19.8–19.9 Potential drop across two scatterers in series calculated from NEGF*

```
1   clear all
2
3   t0=1;Np=51;X=[0:1:Np-1];Nh=floor(Np/2);
4   L=diag([1 zeros(1,Np-1)]);R=diag([zeros(1,Np-1) 1])
        ;
5   zplus=i*1e-12;D=9e-20*t0^2;
6
7   sigB=zeros(Np);siginB=zeros(Np);
8
9   H0=2*t0*diag(ones(1,Np))-t0*diag(ones(1,Np-1),1)-...
10  t0*diag(ones(1,Np-1),-1);
11  N1=Nh-3;N2=Nh+3;UB1=2*t0;UB2=2*t0;
12  H0(N1,N1)=H0(N1,N1)+UB1;H0(N2,N2)=H0(N2,N2)+UB2;H=
        H0;
13
14  EE=0.6*t0;EE=0.81*t0;
15  ck=(1-(EE+zplus)/(2*t0));ka=acos(ck);
16  v=2*t0*sin(ka);
17
18  %Semiclassical profile
19  T=real(v^2/(UB1^2+v^2));R1=(1-T)/T;
20  TT=real(v^2/(D+v^2));R2=0*(1-TT)/TT;
21  RR=[0.5 R2*ones(1,Nh-4) R1 zeros(1,6) R1 R2*ones(1,
        Nh-4) 0.5];
22  RR=cumsum(RR);Vx=ones(1,Np+1)-(RR./RR(Np+1));
23  Fclass=Vx([2:Np+1]);
24  %Based on resistance estimates
```

```
25
26   s1=-t0*exp(i*ka);sig1=kron(L,s1);
27   ck=(1-(EE+zplus)/(2*t0));ka=acos(ck);
28   s2=-t0*exp(i*ka);sig2=kron(R,s2);
29   gam1=i*(sig1-sig1');gam2=i*(sig2-sig2');
30
31   G=inv((EE*eye(Np))-H-sig1-sig2);
32   Tcoh=real(trace(gam1*G*gam2*G'));
33
34   change=100;
35   while change>1e-6
36   G=inv((EE*eye(Np))-H-sig1-sig2-sigB);
37   sigBnew=diag(diag(D*G));sigBnew=D*G;
38   change=sum(sum(abs(sigBnew-sigB)));
39   sigB=sigB+0.25*(sigBnew-sigB);
40   end
41   A=real(diag(i*(G-G')));change=100;
42   while change>1e-6
43   Gn=G*(gam1+siginB)*G';
44   siginBnew=diag(diag(D*Gn));siginBnew=D*Gn;
45   change=sum(sum(abs(siginBnew-siginB)));
46   siginB=siginB+0.25*(siginBnew-siginB);
47   end
48   F=real(diag(Gn))./A;
49
50   hold on
51   h=plot(X,F,'k');
52   set(h,'linewidth',[3.0])
53   %h=plot(X,Fclass,'r-o');
54   %set(h,'linewidth',[1.2])
55   set(gca,'Fontsize',[36])
56   xlabel(' z ---> ')
57   grid on
58   axis([-10 60 0 1])
```

H.2 Chapter 20

H.2.1 *Fig. 20.1 Numerically computed transmission as a function of energy*

```
1  clear all
2
3  %Constants (all MKS, except energy which is in eV)
4  hbar=1.06e-34;q=1.6e-19;qh=q/hbar;B=0;
5
6  %inputs
7  a=2.5e-9;t0=1;
8  NW=25;Np=1;L=zeros(Np);R=L;L(1,1)=1;R(Np,Np)=1;
     zplus=i*1e-12;
9
10 %Hamiltonian
11 al=4*t0;by=-t0;bx=-t0;
12 alpha=kron(eye(NW),al)+kron(diag(ones(1,NW-1),+1),
     by)+kron(diag(ones(1,NW-1),-1),by');
13 alpha=alpha+diag([1:1:NW].*0);
14 alpha=alpha+diag([zeros(1,8) 0*ones(1,9) zeros(1,8)
     ]);
15 beta=kron(diag(exp(i*qh*B*a*a*[1:1:NW])),bx);
16 H=kron(eye(Np),alpha);
17 if Np>1
18 H=H+kron(diag(ones(1,Np-1),+1),beta)+kron(diag(ones
     (1,Np-1),-1),beta');end
19
20 ii=0;for EE=[-0.05:1e-2:1.05]*t0
21 ii=ii+1;ig0=(EE+zplus)*eye(NW)-alpha;
22 if ii==1
23 gs1=inv(ig0);gs2=inv(ig0);end
24 change=1;
25 while change >1e-6
26 Gs=inv(ig0-beta'*gs1*beta);
27 change=sum(sum(abs(Gs-gs1)))/(sum(sum(abs(gs1)+abs(
     Gs))));
28 gs1=0.5*Gs+0.5*gs1;
29 end
```

```
30  sig1=beta'*gs1*beta;sig1=kron(L,sig1);gam1=i*(sig1-
        sig1');
31
32  change=1;
33  while change >1e-6
34  Gs=inv(ig0-beta*gs2*beta');
35  change=sum(sum(abs(Gs-gs2)))/(sum(sum(abs(gs2)+abs(
        Gs))));
36  gs2=0.5*Gs+0.5*gs2;
37  end
38  sig2=beta*gs2*beta';sig2=kron(R,sig2);gam2=i*(sig2-
        sig2');
39
40  G=inv((EE*eye(Np*NW))-H-sig1-sig2);
41  DD=real(diag(i*(G-G')))./2/pi;
42  Tcoh(ii)=real(trace(gam1*G*gam2*G'));E(ii)=EE/t0;ii
43  end
44
45  ii=1;for kk=pi*[-1:0.01:1]
46  H=alpha+beta*exp(i*kk)+beta'*exp(-i*kk);
47  [V,D]=eig(H);EK(:,ii)=sort(abs(diag(D)))./t0;K(ii)=
        kk/pi;ii=ii+1;
48  end
49
50  X=linspace(0,9,101);Ean= 2*(1-cos(pi*X./(NW+1)));
51  hold on
52  figure(1)
53  h=plot(Tcoh,E,'k');
54  set(h,'linewidth',[3.0])
55  %h=plot(X,Ean,'k--');
56  %set(h,'linewidth',[1.2])
57  set(gca,'Fontsize',[36])
58  axis([0 10 -.1 1])
59  % Fig.21.1a, Transmission versus width, at E=t0
60  clear all
61
62  %Constants (all MKS, except energy which is in eV)
63  hbar=1.06e-34;q=1.6e-19;qh=q/hbar;B=0;
```

```
64
65   %inputs
66   a=2.5e-9;t0=1;
67   NW=25;Np=1;L=zeros(Np);R=L;L(1,1)=1;R(Np,Np)=1;
        zplus=i*1e-12;
68
69   %Hamiltonian
70   al=4*t0;by=-t0;bx=-t0;
71   alpha1=kron(eye(NW),al)+kron(diag(ones(1,NW-1),+1),
        by)+kron(diag(ones(1,NW-1),-1),by');
72
73   ii=0;EE=t0*1;for NN=[0:1:NW-1]
74   ii=ii+1;
75
76   alpha=alpha1+diag([zeros(1,NN) 100*ones(1,NW-NN)]);
77   beta=kron(diag(exp(i*qh*B*a*a*[1:1:NW])),bx);
78   H=kron(eye(Np),alpha);
79   if Np>1
80   H=H+kron(diag(ones(1,Np-1),+1),beta)+kron(diag(ones
        (1,Np-1),-1),beta');end
81
82   ig0=(EE+zplus)*eye(NW)-alpha;
83   if ii==1
84   gs1=inv(ig0);gs2=inv(ig0);end
85   change=1;
86   while change >1e-6
87   Gs=inv(ig0-beta'*gs1*beta);
88   change=sum(sum(abs(Gs-gs1)))/(sum(sum(abs(gs1)+abs(
        Gs))));
89   gs1=0.5*Gs+0.5*gs1;
90   end
91   sig1=beta'*gs1*beta;sig1=kron(L,sig1);gam1=i*(sig1-
        sig1');
92
93   change=1;
94   while change >1e-6
95   Gs=inv(ig0-beta*gs2*beta');
```

```
 96  change=sum(sum(abs(Gs-gs2)))/(sum(sum(abs(gs2)+abs(
        Gs)))));
 97  gs2=0.5*Gs+0.5*gs2;
 98  end
 99  sig2=beta*gs2*beta';sig2=kron(R,sig2);gam2=i*(sig2-
        sig2');
100
101  G=inv((EE*eye(Np*NW))-H-sig1-sig2);
102  DD=real(diag(i*(G-G')))./2/pi;
103  Tcoh(ii)=real(trace(gam1*G*gam2*G'));E(ii)=NN;
104  X(ii)=(NN+1)*(acos(1-(EE/2/t0)))/pi;
105  X1(ii)=(NN+1)*sqrt(EE/2/t0)/pi;ii
106
107  end
108
109  hold on
110  figure(1)
111  h=plot(E,Tcoh,'k');
112  set(h,'linewidth',[3.0])
113  %h=plot(E,X,'k');
114  %h=plot(E,X1,'k--');
115  %set(h,'linewidth',[1.2])
116  set(gca,'Fontsize',[36])
117  axis([0 NW -.1 10])
118  grid on
```

H.2.2 *Fig. 20.3 Transmission calculated from NEGF for ballistic graphene sheet and CNT*

```
1  clear all
2
3  %Constants (all MKS, except energy which is in eV)
4  hbar=1.06e-34;q=1.6e-19;qh=q/hbar;a=1e-9;
5
6  %inputs
7
8  t0=-2.5;D=1e-50;ctr=0;zplus=i*1e-3;
9  NL=1;L=zeros(NL);R=L;L(1,1)=1;R(NL,NL)=1;
```

```
10  config=1;%1 for armchair, 2 for zigzag edge
11  NW=floor(14*sqrt(3));% Armchair
12  %NW=14;% Zigzag
13
14  %Hamiltonian
15  al=t0*[0 1 0 0;1 0 1 0;0 1 0 1;0 0 1 0];
16  if config==1
17  bL=t0*[0 0 0 0;0 0 0 0;0 0 0 0;1 0 0 0];
18  bW=t0*[0 0 0 0;1 0 0 0;0 0 0 1;0 0 0 0];end
19  if config==2
20  bW=t0*[0 0 0 0;0 0 0 0;0 0 0 0;1 0 0 0];
21  bL=t0*[0 0 0 0;1 0 0 0;0 0 0 1;0 0 0 0];end
22
23  n=4;% al=4;bW=-1;bL=-1;n=1;
24  alpha=kron(eye(NW),al)+kron(diag(ones(1,NW-1),+1),
        bW)+kron(diag(ones(1,NW-1),-1),bW');
25  alpha=alpha+kron(diag(ones(1,1),1-NW),bW)+kron(diag
        (ones(1,1),NW-1),bW');% for CNT's
26
27  sigB=zeros(NW*NL*n);siginB=zeros(NW*NL*n);
28
29  ii=0;for EE=t0*[-0.5:+0.01:+0.5]
30  ii=ii+1;
31  ig0=(EE+zplus)*eye(NW*n)-alpha;
32  if ii==1
33  gs1=inv(ig0);gs2=inv(ig0);end
34
35  BB=0;beta=kron(diag(exp(i*qh*BB*a*a*[1:1:NW])),bL);
36  %beta=kron(eye(NW),bL);
37  H=kron(eye(NL),alpha);if NL>1
38  H=H+kron(diag(ones(1,NL-1),+1),beta)+
39  kron(diag(ones(1,NL-1),-1),beta');end
40
41  change=1;
42  while change>1e-4
43  Gs=inv(ig0-beta'*gs1*beta);
44  change=sum(sum(abs(Gs-gs1)))/(sum(sum(abs(gs1)+abs(
        Gs))));
```

```
45  gs1=0.5*Gs+0.5*gs1;
46  end
47  sig1=beta'*gs1*beta;sig1=kron(L,sig1);gam1=i*(sig1-
        sig1');
48  change=1;
49  while change>1e-4
50  Gs=inv(ig0-beta*gs2*beta');
51  change=sum(sum(abs(Gs-gs2)))/(sum(sum(abs(gs2)+abs(
        Gs))));
52  gs2=0.5*Gs+0.5*gs2;
53  end
54  sig2=beta*gs2*beta';sig2=kron(R,sig2);gam2=i*(sig2-
        sig2');
55
56  G=inv((EE*eye(NW*NL*n))-H-sig1-sig2);
57  T(ii)=real(trace(gam1*G*gam2*G'));E(ii)=EE/t0;
58  if EE==0
59  T(ii)=T(ii-1);end,EE
60  end
61
62
63  hold on
64  h=plot(T,E,'k');
65  set(h,'linewidth',[3.0])
66  set(gca,'Fontsize',[36]);
67  axis([0 10 -0.5 +0.5])
68  title(' W = 24 * 2b ')
69  grid on
```

H.2.3 *Fig. 20.4 Normalized Hall resistance versus B-field for ballistic channel*

```
1  clear all
2
3  %Constants (all MKS, except energy which is in eV)
4  hbar=1.06e-34;q=1.6e-19;m=0.1*9.1e-31;qh=q/hbar;
5
6  %inputs
```

```
a=2.5e-9;t0=(hbar^2)/(2*m*(a^2)*q);
NW=25;Np=1;L=zeros(Np);R=L;L(1,1)=1;R(Np,Np)=1;
    zplus=i*1e-12;

%Hamiltonian
al=4*t0;by=-t0;bx=-t0;
alpha=kron(eye(NW),al)+kron(diag(ones(1,NW-        1)
    ,+1),by)+kron(diag(ones(1,NW-1),-1),by');
alpha=alpha+diag([1:1:NW].*0);

EE=t0;ii=0;for B=0:0.1:50
%B=0;ii=0;for EE=[-0.05:0.01:1]*t0
ii=ii+1;E(ii)=B;
ig0=(EE+zplus)*eye(NW)-alpha;
if ii==1
gs1=inv(ig0);gs2=inv(ig0);end

beta=kron(diag(exp(i*qh*B*a*a*[1:1:NW])),bx);
H=kron(eye(Np),alpha);
if Np>1
H=H+kron(diag(ones(1,Np-1),+1),beta)+kron(diag(ones
    (1,Np-1),-1),beta');end

change=1;
while change >5e-5
Gs=inv(ig0-beta'*gs1*beta);
change=sum(sum(abs(Gs-gs1)))/(sum(sum(abs(gs1)+abs(
    Gs))));
gs1=0.5*Gs+0.5*gs1;
end
sig1=beta'*gs1*beta;sig1=kron(L,sig1);gam1=i*(sig1-
    sig1');

change=1;
while change >5e-5
Gs=inv(ig0-beta*gs2*beta');
change=sum(sum(abs(Gs-gs2)))/(sum(sum(abs(gs2)+abs(
    Gs))));
```

```
39  gs2=0.5*Gs+0.5*gs2;
40  end
41  sig2=beta*gs2*beta';sig2=kron(R,sig2);gam2=i*(sig2-
        sig2');
42
43  G=inv((EE*eye(Np*NW))-H-sig1-sig2);
44  Gn=G*gam1*G';
45
46  A=i*(G-G');V=real(diag(Gn./A));
47  Tcoh=real(trace(gam1*G*gam2*G'));TM=real(trace(gam2
        *Gn));
48  %Y(ii)=Tcoh;ii
49  Y(ii)=(V(1)-V(NW))/Tcoh;ii
50  end
51
52  hold on
53  h=plot(E,Y,'k');
54  set(h,'linewidth',[3.0])
55  set(gca,'Fontsize',[36])
56  xlabel(' B-field (T) ---> ')
57  ylabel(' R_{xy} ---> ')
58  grid on
```

H.2.4 *Fig. 20.5 Grayscale plot of local density of states*

```
1   clear all
2
3   %Constants (all MKS, except energy which is in eV)
4   hbar=1.06e-34;q=1.6e-19;m=0.1*9.1e-31;qh=q/hbar;B
        =20;
5
6   %inputs
7   a=2.5e-9;t0=(hbar^2)/(2*m*(a^2)*q);
8   NW=25;Np=1;L=zeros(Np);R=L;L(1,1)=1;R(Np,Np)=1;
        zplus=i*1e-12;
9
10  %Hamiltonian
11  al=4*t0;by=-t0;bx=-t0;
```

```
12  alpha=kron(eye(NW),al)+kron(diag(ones(1,NW-1),+1),
       by)+kron(diag(ones(1,NW-1),-1),by'));
13  alpha=alpha+diag([1:1:NW].*0);
14  beta=kron(diag(exp(i*qh*B*a*a*[1:1:NW])),bx);
15  H=kron(eye(Np),alpha);
16  if Np>1
17  H=H+kron(diag(ones(1,Np-1),+1),beta)+kron(diag(ones
       (1,Np-1),-   1),beta');end
18
19  ii=0;for EE=[-0.05:0.008:1.05]*t0
20  ii=ii+1;ig0=(EE+zplus)*eye(NW)-alpha;
21  if ii==1
22  gs1=inv(ig0);gs2=inv(ig0);end
23  change=1;
24  while change >1e-4
25  Gs=inv(ig0-beta'*gs1*beta);
26  change=sum(sum(abs(Gs-gs1)))/(sum(sum(abs(gs1)+abs(
       Gs))));
27  gs1=0.5*Gs+0.5*gs1;
28  end
29  sig1=beta'*gs1*beta;sig1=kron(L,sig1);gam1=i*(sig1-
       sig1');
30
31  change=1;
32  while change >1e-4
33  Gs=inv(ig0-beta*gs2*beta');
34  change=sum(sum(abs(Gs-gs2)))/(sum(sum(abs(gs2)+abs(
       Gs))));
35  gs2=0.5*Gs+0.5*gs2;
36  end
37  sig2=beta*gs2*beta';sig2=kron(R,sig2);gam2=i*(sig2-
       sig2');
38
39  G=inv((EE*eye(Np*NW))-H-sig1-sig2);
40  DD(:,ii)=real(diag(i*(G-G'))). /2/pi;
41  Gn=G*gam1*G';
42  NN(:,ii)=real(diag(Gn)). /2/pi;
43  end
```

```
44
45  XX=DD;
46  lo=.4*min(min(XX));hi=.4*max(max(XX));
47
48  figure(1)
49  hold on
50  imagesc(XX,[lo hi])
51  colormap(gray)
52  set(gca,'Fontsize',[36])
53  grid on
54  axis([0 140 0 25])
```

H.3   Chapter 22

H.3.1   *Fig. 22.7, n versus μ, single dot*

```
1   clear all
2   %define constants
3   eps = 10; U = 20;
4
5   %define N and H matrices
6   N = diag([0 1 1 2]);
7   H = diag([0 eps eps 2*eps+U]);
8
9   ii=1;for mu = 0:0.1:50
10  p = expm(-(H-mu*N));
11  rho = p/trace(p);
12  n(ii) = trace(rho*N);X(ii)=mu;ii=ii+1;
13  end
14
15  G=diff(n);G=[0 G];G=G./max(G);
16  hold on;
17  grid on;
18  h=plot(X,n,'k');
19  set(h,'linewidth',2.0)
20  set(gca,'Fontsize',36)
21  xlabel('\mu / kT --->');
22  ylabel(' n   ---> ');
```

H.3.2   *Fig. 22.8, I versus V, single quantum dot*

```
1   clear all
2   %define constants
3   eps = 10; U = 20; g1=1; g2=1;N = [0 1 1 2];
4
5   ii=1;mu1=0;for mu2=0:1:50
6   f1a=g1/(1+exp(eps-mu1));
7   f2a=g2/(1+exp(eps-mu2));
8   f1b=g1/(1+exp(eps+U-mu1));
9   f2b=g2/(1+exp(eps+U-mu2));
10
11  W1=[0 g1-f1a g1-f1a 0;
12  f1a 0 0 g1-f1b;
13  f1a 0 0 g1-f1b;
14  0 f1b f1b 0];W1=W1-diag(sum(W1));
15
16  W2=[0 g2-f2a g2-f2a 0;
17  f2a 0 0 g2-f2b;
18  f2a 0 0 g2-f2b;
19  0 f2b f2b 0];W2=W2-diag(sum(W2));
20
21  W=W1+W2;
22  [V,D]=eig(W);diag(D);
23  P=V(:,1);P=P./sum(P);
24  I1(ii)=N*W1*P;
25  I2(ii)=N*W2*P;
26  X(ii)=mu2;ii=ii+1;
27  end
28
29  grid on;
30  %h=plot(X,I1,'k');
31  %set(h,'linewidth',2.0)
32  h=plot(X,I2,'k');
33  set(h,'linewidth',2.0)
34  set(gca,'Fontsize',36)
35  xlabel('qV/kT --->');
36  ylabel(' Normalized current ---> ');
```

H.3.3   *Fig. 22.9, n versus μ, double quantum dot*

```
clear all
%define constants
eps1 = 20; eps2 = 20; t = 10;
U = 20;

%define N and H matrices
N = diag([ones(1,1)*0 ones(1,4)*1 ones(1,6)*2 ones
    (1,4)*3 ones(1,1)*4]);
H0 = 0;
h11 = [eps1 t;t eps2];H1 = blkdiag(h11,h11);
h21 = [2*eps1+U 0;0 2*eps2+U];
h22 = [eps1+eps2 0;0 eps1+eps2];
H2 = blkdiag(h21,h22,h22);H2(1:2,3:4)=t;H2(3:4,1:2)
    =t;
h31=[eps1+2*eps2+U t;t 2*eps1+eps2+U];H3 = blkdiag(
    h31,h31);
H4 = 2*eps1+2*eps2+2*U;
H = blkdiag(H0,H1,H2,H3,H4);

ii=1;for mu = 0:60
p = expm(-(H-mu*N));
rho = p/trace(p);
n(ii) = trace(rho*N);X(ii)=mu;ii=ii+1;
end

hold on;
grid on;
box on;
h=plot(X,n,'k');
set(h,'linewidth',2.0)
set(gca,'Fontsize',36)
xlabel('\mu / kT --->');
ylabel(' n ---> ');
```

H.4 Chapter 23

H.4.1 *Fig. 23.9 Voltage probe signal as the magnetization of the probe is rotated*

```
1  clear all
2  hbar=1.06e-34;q=1.6e-19;m=0.1*9.1e-31;a=2.5e-9;t0=(
      hbar^2)/(2*m*(a^2)*q);
3  sx=[0 1;1 0];sy=[0 -i;i 0];sz=[1 0;0 -1];zplus=1i*1
      e-12;
4
5  Np=50;N1=10;N2=20;X=1*[0:1:Np-1];
6  L=diag([1 zeros(1,Np-1)]);
7  R=diag([zeros(1,Np-1) 1]);
8  L1=0.1*diag([zeros(1,N1-1) 1 zeros(1,Np-N1)]);
9  L2=0.1*diag([zeros(1,N2-1) 1 zeros(1,Np-N2)]);
10
11 ii=0;for theta=[0:0.1:4]*pi
12 P1=0.7*[0 0 1];
13 P2=1*[sin(theta) 0 cos(theta)];ii=ii+1;
14
15 H0=diag(ones(1,Np));
16 HR=diag(ones(1,Np-1),1);HL=diag(ones(1,Np-1),-1);
17
18 H=2*t0*kron(H0,eye(2))-t0*kron(HL,eye(2))-t0*kron(
      HR,eye(2));
19
20 EE=t0;ck=(1-(EE+zplus)/(2*t0));ka=acos(ck);
21 sL=-t0*exp(1i*ka)*eye(2);sR=sL;
22 s1=-t0*exp(1i*ka)*(eye(2)+P1(1)*sx+P1(2)*sy+P1(3)*
      sz);
23 s2=-t0*exp(1i*ka)*(eye(2)+P2(1)*sx+P2(2)*sy+P2(3)*
      sz);
24
25 sigL=kron(L,sL);sigR=kron(R,sR);
26 sig1=kron(L1,s1);sig2=kron(L2,s2);
27 gamL=1i*(sigL-sigL');gamR=1i*(sigR-sigR');
28 gam1=1i*(sig1-sig1');gam2=1i*(sig2-sig2');
29
```

```
30  G=inv(((EE+zplus)*eye(2*Np))-H-sigL-sigR-sig1-sig2)
       ;
31
32  % {1 L} {2 R} = {a} {b}
33
34  TM1L=real(trace(gam1*G*gamL*G'));
35  TML1=real(trace(gamL*G*gam1*G'));
36  Taa=[0 TM1L;TML1 0];
37
38  TM12=real(trace(gam1*G*gam2*G'));
39  TM1R=real(trace(gam1*G*gamR*G'));
40  TML2=real(trace(gamL*G*gam2*G'));
41  TMLR=real(trace(gamL*G*gamR*G'));
42  Tab=[TM12 TM1R;TML2 TMLR];
43
44  TM21=real(trace(gam2*G*gam1*G'));
45  TM2L=real(trace(gam2*G*gamL*G'));
46  TMR1=real(trace(gamR*G*gam1*G'));
47  TMRL=real(trace(gamR*G*gamL*G'));
48  Tba=[TM21 TM2L;TMR1 TMRL];
49
50  TM2R=real(trace(gam2*G*gamR*G'));
51  TMR2=real(trace(gamR*G*gam2*G'));
52  Tbb=[0 TM2R;TMR2 0];
53
54  Taa=diag(sum(Taa)+sum(Tba))-Taa;Tba=-Tba;
55  Tbb=diag(sum(Tab)+sum(Tbb))-Tbb;Tab=-Tab;
56  if abs(sum(sum([Taa Tab;Tba Tbb]))) > 1e-10
57  junk=100,end
58
59  V=-inv(Tbb)*Tba*[1;0];
60  VV2(ii)=V(1);VVR(ii)=V(2);
61  angle(ii)=theta/pi;
62  I2(ii)=TM21;IL(ii)=TML1;IR(ii)=TMR1;
63
64  Gn=G*(gam1+V(1)*gam2+V(2)*gamR)*G';
65  Gn=Gn([2*N2-1:2*N2],[2*N2-1:2*N2]);
66  A=i*(G-G');A=A([2*N2-1:2*N2],[2*N2-1:2*N2]);
```

```
67  g2=i*(s2-s2');XX2(ii)=real(trace(g2*Gn)/trace(g2*A)
        );
68  end
69  X2=VV2;XR=VVR;max(X2)-min(X2)
70
71  hold on
72  h=plot(angle,X2,'k');
73  set(h,'linewidth',2.0)
74  h=plot(angle,XX2,'ro');
75  set(h,'linewidth',2.0)
76  set(gca,'Fontsize',36)
77  xlabel(' \theta / \pi ---> ')
78  ylabel(' V_{2} ---> ')
79  grid on
```

H.4.2   *Fig. 23.10 Voltage probe signal due to variation of gate voltage controlled Rashba coefficient*

```
1   clear all
2   hbar=1.06e-34;q=1.6e-19;m=0.1*9.1e-31;a=2.5e-9;t0=(
        hbar^2)/(2*m*(a^2)*q);
3   sx=[0 1;1 0];sy=[0 -1i;1i 0];sz=[1 0;0 -1];zplus=1i
        *1e-12;
4
5   Np=50;N1=5;N2=45;X=1*[0:1:Np-1];
6   L=diag([1 zeros(1,Np-1)]);
7   R=diag([zeros(1,Np-1) 1]);
8
9   L1=0.1*diag([zeros(1,N1-1) 1 zeros(1,Np-N1)]);
10  L2=0.1*diag([zeros(1,N2-1) 1 zeros(1,Np-N2)]);
11
12  ii=0;for al=[0:0.005:0.3]*t0
13  P1=[0 0 0.7];P2=[0 0 0.7];ii=ii+1;
14  alph=al*1;% Rashba
15  BB=al*0;% Hanle
16
17  H0=diag(ones(1,Np));
18  HR=diag(ones(1,Np-1),1);HL=diag(ones(1,Np-1),-1);
```

```
19  beta=t0*eye(2)+1*i*alph*sx;
20  alpha=2*t0*eye(2)+1*BB*sx;
21
22  H=kron(H0,alpha)-kron(HL,beta')-kron(HR,beta);
23
24  EE=t0;ck=(1-(EE+zplus)/(2*t0));ka=acos(ck);
25  sL=-t0*exp(1i*ka)*eye(2);sR=sL;
26  s1=-t0*exp(1i*ka)*(eye(2)+P1(1)*sx+P1(2)*sy+P1(3)*
      sz);
27  s2=-t0*exp(1i*ka)*(eye(2)+P2(1)*sx+P2(2)*sy+P2(3)*
      sz);
28
29  sigL=kron(L,sL);sigR=kron(R,sR);
30  sig1=kron(L1,s1);sig2=kron(L2,s2);
31  gamL=1i*(sigL-sigL');gamR=1i*(sigR-sigR');
32  gam1=1i*(sig1-sig1');gam2=1i*(sig2-sig2');
33
34  G=inv(((EE+zplus)*eye(2*Np))-H-sigL-sigR-sig1-sig2)
      ;
35
36  % {1 L} {2 R} = {a} {b}
37  TM1L=real(trace(gam1*G*gamL*G'));
38  TML1=real(trace(gamL*G*gam1*G'));
39  Taa=[0 TM1L;TML1 0];
40
41  TM12=real(trace(gam1*G*gam2*G'));
42  TM1R=real(trace(gam1*G*gamR*G'));
43  TML2=real(trace(gamL*G*gam2*G'));
44  TMLR=real(trace(gamL*G*gamR*G'));
45  Tab=[TM12 TM1R;TML2 TMLR];
46
47  TM21=real(trace(gam2*G*gam1*G'));
48  TM2L=real(trace(gam2*G*gamL*G'));
49  TMR1=real(trace(gamR*G*gam1*G'));
50  TMRL=real(trace(gamR*G*gamL*G'));
51  Tba=[TM21 TM2L;TMR1 TMRL];
52
53  TM2R=real(trace(gam2*G*gamR*G'));
```

```
54  TMR2=real(trace(gamR*G*gam2*G'));
55  Tbb=[0 TM2R;TMR2 0];
56
57  Taa=diag(sum(Taa)+sum(Tba))-Taa;Tba=-Tba;
58  Tbb=diag(sum(Tab)+sum(Tbb))-Tbb;Tab=-Tab;
59  if abs(sum(sum([Taa Tab;Tba Tbb]))) > 1e-10
60  junk=100,end
61
62  V=-inv(Tbb)*Tba*[1;0];
63  VV2(ii)=V(1);VVR(ii)=V(2);
64  alp(ii)=al*2/t0;% eta/t0/a
65
66  Gn=G*(gam1+V(1)*gam2+V(2)*gamR)*G';A=i*(G-G');
67  VV(ii)=real(trace(gam2*Gn)/trace(gam2*A));
68  end
69
70  hold on
71  h=plot(alp,VV2,'k');
72  set(h,'linewidth',2.0)
73  h=plot(alp,VV,'ro');
74  set(h,'linewidth',2.0)
75  set(gca,'Fontsize',36)
76  ylabel(' Voltage ---> ')
77  grid on
```

Appendix I

Table of Contents of Part A: Basic Concepts

Appendix J

Available Video Lectures for Part A: Basic Concepts

Following is a detailed list of *video lectures* available at the course website corresponding to different sections of this volume (Part A: Basic Concepts).

Lecture in Course website	Topic	Discussed in this book
Scientific Overview	*Overview*	Chapter 1
Unit 1: L1.1	*Introduction*	Chapters 2–4
Unit 1: L1.2	*Two Key Concepts*	Section 2.1
Unit 1: L1.3	*Why Electrons Flow*	Chapter 3
Unit 1: L1.4	*Conductance Formula*	Chapter 3
Unit 1: L1.5	*Ballistic* (B) *Conductance*	Chapter 4
Unit 1: L1.6	*Diffusive* (D) *Conductance*	Chapter 4
Unit 1: L1.7	*Ballistic* (B) *to Diffusive* (D)	Chapter 4
Unit 1: L1.8	*Angular Averaging*	Ch. 4, App. B
Unit 1: L1.9	*Drude Formula*	Section 2.5.1
Unit 1: L1.10	*Summary*	Chapters 2–4
Unit 2: L2.1	*Introduction*	Chapter 6
Unit 2: L2.2	$E(p)$ *or* $E(k)$ *Relation*	Section 6.2
Unit 2: L2.3	*Counting States*	Section 6.3
Unit 2: L2.4	*Density of States*	Section 6.3.1
Unit 2: L2.5	*Number of Modes*	Section 6.4
Unit 2: L2.6	*Electron Density* (n)	Section 6.5
Unit 2: L2.7	*Conductivity vs.* n	Section 6.6
Unit 2: L2.8	*Quantum Capacitance*	Section 7.3.1

Index